Praise for *Animal Architects*

"What is special—and admirable—about James and Carol Gould's *Animal Architects* is their focus on built constructions. . . . *Animal Architects* covers a lot of ground with a minimum of jargon and a temperate, nondidactic tone. What may be most attractive about *Animal Architects* is the way it takes a series of examples and makes them work as elements of a general story, showing how natural selection manifests itself in the real world, not least in the form of nonverbal intelligence, something we can't easily envision."

—*Wall Street Journal*

"The Goulds have written one of the most provocative books I have ever read. *Animal Architects* explores one of the most interesting scientific questions: whether the ability of insects, birds and mammals to build complex structures is what we humans would consider intelligent behavior. The Goulds' answer is clear: Many creatures possess what humans would traditionally define as intelligence."

—*Roanoke Times*

"One of the best popular science books of recent years."

—*The Spectator* (UK)

"In *Animal Architects*, James R. Gould and Carol Grant Gould beautifully describe some of the architectural wonders of the animal kingdom. . . . There is no doubt that the Goulds succeed in . . . captivating the reader with their enthusiasm and encyclopaedic knowledge of the biology of building."

—*Times Literary Supplement*

"The story of this amazing and beautifully written little book is one of humans very gradually, and only through gritted teeth, admitting that other animals, down to the apparently humblest insects, are more intelligent than was ever suspected."

—*Guardian* (UK)

"Raises interesting questions on animal intelligence and the evolution of building behavior."

—*Library Journal*

"A fascinating journey, with plenty of surprises. . . . This book is filled with fascinating vignettes illuminating the intelligence capabilities of species us humans would like to think of as inferior; again and again, the Goulds show that human beings aren't necessarily the smartest kids in class."

—*Publishers Weekly*

"'Miracle' is not a word often found in reference to science, yet to the lay reader, this book will seem miraculous. For me, after a lifetime of reading and hands-on experience concerning the natural world, this spectacular book came as an explosion of surprising facts and interrelated studies defining intricacies so profound that one's only desire is to learn more. If asked to choose a single book that shows how the natural world is not merely a landscape, a single book that conveys the value of science and demands respect for the complex lives that are lived all around us, I'd have to choose this one."

—Elizabeth Marshall Thomas, author of
The Hidden Life of Dogs and
The Old Way: A Story of the First People

"Nature is full of great architects, from termites that build air-conditioned skyscrapers to male bowerbirds, who decorate their nests with such a 'sense of beauty' that no female can resist. James and Carol Gould leave no stone unturned in their delightful description of these marvels. They puzzle, with the reader, what kind of interplay between inborn concepts and experience guides their production."

—Frans de Waal, author of *Our Inner Ape*

"*Animal Architects* is a fascinating demonstration of the essential continuity between humans and animals—as builders, and as intelligent beings."

—Peter Singer, Professor of Bioethics,
Princeton University, and author of *Animal Liberation*

"In this forward-looking and exciting book, Jim and Carol Grant Gould open the door for informed discussion of intelligence in animals frequently written off as unintelligent automatons and force us to revise our stereotypes about insect and other animal architects."

—Marc Bekoff, author of
The Emotional Lives of Animals

ANIMAL ARCHITECTS

Building *and the* Evolution *of* Intelligence

JAMES L. GOULD *and*
CAROL GRANT GOULD

BASIC BOOKS
A Member of the Perseus Books Group
New York

Hardcover first published in 2007 by Basic Books
A Member of the Perseus Books Group
Paperback first published in 2012 by Basic Books

Designed by Brent Wilcox

The Library of Congress has catalogued the hardcover edition as follows:
Gould, James L., 1945–
    Animal architects : building and the evolution of intelligence / James L. Gould, Carol Grant Gould.
        p. cm.
     ISBN-13: 978-0-465-02782-8
     ISBN-10: 0-465-02782-2
     1. Animals—Habitations.   2. Animal intelligence.   I. Gould, Carol Grant.
II. Title.
QL756.G57 2007
591.5—dc22

                                                        2006036042

ISBN: 978-0-465-02838-2 (paperback)
ISBN: 978-0-465-02839-9 (e-book)

10 9 8 7 6 5 4 3 2 1

To Doris Mae Gould
with love

CONTENTS

PREFACE

THE ENGINEERING SKILL that goes into a beaver dam rivals the elegant calculations that built the pyramids and the Panama Canal. Furry rodents with paddle tails and oversized incisors gauge flow rates and stress, fell trees and cut them to size, and build roads and excavate canals to float the logs and branches to the construction site. They erect custom designed multipurpose underwater bunkers and keep the water level just right, adjusting the flow and stopping leaks without the aid of opposable thumbs or even fingers. In the winter, when ice covers their ponds, they drain off just enough to make a breathing space beneath the ice ceiling.

Given their size and lineage, beavers put humans to shame when it comes to architecture. Their accomplishments raise issues of planning and awareness that are quite new to scientific discourse. And it doesn't stop with the beavers: animals of many orders, from primates to arthropods, build homes, traps, climate-control systems, civil engineering projects, fashion-show runways, and nurseries out of paper, silk, adobe, wax, clay, sticks, grass, fibers, or lichen. Some of their construction is guided wholly by instinct, some benefits from practice, some even suggests insight and, in some instances, a kind of innovation that seems to require the understanding needed to deal with unforeseen situations and problems.

Three nests of a potter wasp.

The goal of this book is to appreciate the diversity of animal artifacts, and to understand the general strategies that allow creatures from insects through mammals to create the structures that characterize their species. We hope to discover how these architectural marvels solve the ecological problems posed by each animal's habitat and niche, and we will look at the ways decision making and apparent planning play a role in building and the ongoing evolution of species. We will discover that construction behavior in turn impacts both the mind and the niche, driving behavioral evolution. And we will also see how totally different our own building is—how humans, with the highest ratio of brain to body on the planet, are at a loss to reproduce the architectural feats of nearly any other animal.

Our choice of examples is largely governed by the need to compare innate and learned components of building, and to look for suggestions of planning, even aesthetic "taste." Revealing experiments are at a premium since the possibility of some combination of instinct and experience, even cognitive processing, edged into the intellectual Zeitgeist of most students of behavior only very recently. As a result, in many instances we have had to rely on intuition and analogy to analyze and infer what is going on in the minds of animal architects. The result is a blend of openness and skepticism that we hope will generate and renew interest in animal building and the neural equipment that makes these remarkable achievements possible.

Jim and Carol Gould
Princeton, New Jersey

CHAPTER I

Why Animals Build

WE ARE ACCUSTOMED to thinking of humans as the world's master builders. Look around you: chances are that essentially everything you see will have been made by humans. Walls or floors, furniture or books, cathedrals or cell phone towers, our species appears to dominate the market for housing and artifacts. This impression is reinforced away from civilization; in a forest, on a plain, or in a desert, little or nothing of what we see is made by nonhuman animals. Trees and grasses are vegetative products; unstructured sand and stone dominate barren habitats.

As a species whose ancestors were probably living in the forest canopy as recently as 15 million years ago, we have come a long way. Our immense cognitive capacity and language make it possible to hand down discoveries from generation to generation, and so we build upon existing knowledge.

And yet the largest structures built by animals are not human creations at all. A quick look at Google Earth will convince anyone that (apart from smog and areas of green) the only indications of life on our planet visible from space are coral reefs. The Great Barrier Reef off Australia is about a thousand miles long, and it is evident to the naked eye from about ten thousand miles away. Reefs are doubly impressive because the builders are so tiny: the coral animals, standing on the calcareous platforms they build for themselves, are a fraction of an inch in diameter. Much of what goes into plastic and gasoline is thought to have been once locked up in coral reefs; the remains of these minute, long-dead animal builders thus make modern life possible.

Another instructive comparison: our tallest buildings reach less than three-tenths of a mile into the air, and our deepest wells seldom penetrate more than five miles. Individual termites are less than a tenth of an inch long, yet they build towers twenty feet high; at the human scale, this is the equivalent of two and a half miles. To reach water they dig wells up to 150 feet deep, the equivalent of nearly twenty miles for humans; individual workers may walk that distance up and down many times each day.

The termite projects—towers and wells, ventilation shafts, nurseries, feeding runways, and the complex suspended chamber that encloses the king and queen—require monumental coöperation. But this smooth-running society comprises tens of thousands of sightless individuals that never seem to need instructions. How is it possible that these blind, unprepossessing insects manage to build their castles of clay? What cognitive tools do they bring to the task? Which intellectual abilities might we have in common, and how is it that they play out so differently from one species to another?

Why Build?

The needs of a species depend on its niche—its way of making a living in a hostile and competitive world. The degree of neural sophistication available to an animal ought to be related to the challenge of its niche, and to the complexity of the brain that has evolved to solve the problems the species faces.

To begin to understand what goes on when animals build, we need to be aware of the challenges the creatures are trying to overcome. We must know what resources and limitations they have to deal with, and what onboard equipment evolution has provided. To grasp why animals build and what mental capacities are at their disposal, we have to recognize the possibility that animals might use some degree of creativity or insight in their construction projects.

From a biological point of view, life is about survival and reproduction. This means finding and capturing food; avoiding an early death from predation, exposure to the elements, or chance catastrophe; finding and attracting mates; and, for some species, protecting and caring for the young. Where the ability to build has evolved, it should be in aid of these needs. Spears, guns, knives and machetes, traps, nets, and fish hooks are some of the many weapons humans have fashioned for the killing or collecting of food. We've devised baskets and bags for transporting, containers for storing, pans for cooking, and implements for eating. With no tools or other artifacts, with only our hands, how could we survive? Animals have separately invented most of these tools; they may be born with them preinstalled as specialized body parts, or must create them from everyday materials. But no species has managed anything like the diversity of tools seen in humans.

To survive as a species, we limit our exposure and risk in many ways. Clothes allow humans to live in temperate and even arctic climates, places where for part of the year we would otherwise die from heat loss. Polar bears and lions, on the other hand, cannot change clothes, and they have correspondingly limited ranges. Our houses keep us dry, and they moderate the outside extremes of temperature, wind, and humidity. Predation, unless at the hands of one another, is not a factor in the human niche, but it is a major concern for other animals.

We are among the minority of species that employ artifacts to attract mates. Humans use clothes, cosmetics, housing, and material goods to serve as evidence to potential mates of their health, wealth, and ability to compete; the artificial enhancement of such cues is a tradition of at least a hundred thousand years' standing. But the human use of decoration and advertising propaganda pales in comparison to that of some birds, as we shall see.

In our species, caring for the young generally involves the same kinds of clothing, housing, and tools used by adults, but many of the artifacts animals build are designed exclusively for protecting their offspring. Given the much greater risk from predation and exposure the young face, this should be no great surprise. For nonhuman animals, the chance of dying without issue in the struggle to survive and reproduce is very high.

Just as life as we know it would be impossible without the artifacts we make, many species have no choice but to modify their environments in analogous ways to survive. The lifestyle of the species dictates the kinds of tools or edifices they need. How they do it is the question common to them all. What do the individuals of a particular species know innately, and how do they organize and use such information? How do they know what to learn and when they are making progress with a problem? And when do animals need to understand what they are about?

INSTINCTIVE RESPONSES

Given the enormous complexity of bird nests and beaver dams, early authorities on animal behavior assumed that elaborate construction behavior must be learned. But naturalists never really bought into this explanation; they observed nests being built by inexperienced first-timers and saw that, despite imperfections, they had most of the characteristics seen in the constructions of mature adults. How this could be remained for centuries an awesome mystery of nature. Until comparatively recently, neither the neural mechanisms of instinct nor the subtle and arcane rules of conditioning were understood. And only the most unreserved romantics conceived of a role for planning and thought in any brain but ours. A hundred years ago, students of animal behavior thought in simplistic either/or terms: an animal did this or that using either instinct or learning. We now understand that much behavior draws liberally on both strategies, and others not then thought of.

We know now that animal building, like most tasks animals accomplish, depends on many different neural mechanisms. For most creatures, though, instinct rather than learning seems to guide behavior. Early ethologists—biologists who specialize in animal behavior—were able to puzzle out the workings of instinct from a few closely analyzed animal systems. The studies of Konrad Lorenz and Niko Tinbergen on greylag geese are outstanding and illuminating examples. (Lorenz and Tinbergen later shared a Nobel Prize with Karl von Frisch for their pioneering studies.)

Geese build nests on the ground; they are made of sticks and twigs and lined with down the parents pluck from their breasts. The down insulates the eggs, and the skin thus exposed serves as a brood patch that the birds press against the eggs to warm them. Lorenz and Tinbergen did not study the building behavior itself in much detail, but were struck that courtship, nest building, egg laying, incubation, and the rearing of young occurred in a tightly coördinated annual cycle. This cycle required that the birds initiate behaviors in advance of need or overt cues—to build a nest before there were eggs, for instance. Surely, they thought, there must be a built-in drive in place, but suppressed, that allowed behaviors to appear when needed. The correct timing and sequence simply could not be worked out by wasteful trial and error, bad timing, and wrong guesses, each of which would endanger the all-important next generation.

Lorenz and Tinbergen were particularly drawn to one curious behavior. Most birds turn their eggs once or twice a day, which we now know prevents the embryo from sticking to the inside of the shell. Many ground-nesting birds build nests with shallow sides, and

Egg rolling. When an incubating goose spots an egg-like object near the nest, it extends its neck forward, rises, and places its bill just beyond the object. The goose then rolls the egg back between its legs into the nest. Once this recovery sequence begins, it continues to completion even if the egg is removed.

thus risk the loss of an occasional egg over the side during the turning ritual. When a bird is on the nest, it notices a nearby egg and recovers it, rolling it carefully up and over the side and back in. The process of rolling the egg back into the nest looks at first like an intelligent and thoughtful process: a problem has been diagnosed, a solution found and carried out.

The trouble is that all greylags recover errant eggs in the same way. When a goose spots an egg outside the nest, it fixes its gaze on the egg, rises, steps forward, puts its bill on top of the egg, and then rolls it back between its legs. No goose, even on its first attempt, ever tries to push the egg, or to use a foot or a wing, or to roll the egg from the side. Most other ground-nesting birds recover eggs in this one stereotyped way. A "thoughtful" problem-solving approach would be bound to produce not only alternative solutions but also the inevitable errors that come with experimentation.

To Lorenz and Tinbergen, the process of egg recovery looked totally mindless and automatic. Because some of the birds they observed had been hand reared, they were able to get close enough to the nests to confirm this impression. Even when they removed an egg after the goose had risen to its feet and extended its neck, the entire egg-rolling response unfolded regardless, right down to the careful tucking and settling motions that nestled an imaginary egg securely beneath its parent. It is clear that a preëxisting neural circuit orchestrates the operation of the dozens of muscles involved in recovering the egg; it controls the timing and relative contraction strength of muscles that, in other contexts, are used for calling, swallowing, feeding, courtship and other signaling, walking, and so on. Once triggered, the behavior progresses unconsciously.

Lorenz and Tinbergen called the behavior that this presumed circuit produced a fixed-action pattern; today it is more often called a *motor program*. Neurobiologists have mapped many circuits controlling stereotyped behavior, cell by cell, and virtually all innate actions involving muscle coördination are controlled in this way. Motor programs are an invaluable shortcut in building behavior.

INSTINCTIVE RECOGNITION

But the recovery response is triggered by and directed at eggs. How do geese recognize eggs, their own all-important units of reproductive fitness, in the first place? We might suppose that having incubated the eggs for up to three weeks, carefully turning them from time to time, the parents would have learned their appearance. But although this does happen in a few unusual species, it is not typical. We can place a motley variety of objects around a nest that the parents will

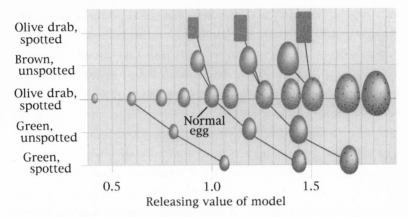

Olive drab, spotted
Brown, unspotted
Olive drab, spotted
Green, unspotted
Green, spotted

Normal egg

0.5   1.0   1.5

Releasing value of model

Eggness test. Baerends offered nesting geese a pair of egg-like objects to choose between and scored which alternative was rolled first. From a thorough series of such choices, he was able to calculate the relative effectiveness of each model. Green is the best color, spotting is important, increasing size makes the object more attractive, and an ovoid shape is better than a rectangular solid.

methodically roll in: light bulbs, beer bottles, grapefruit—almost anything not too dense that is three-dimensional and has a continuous convex edge. Fortunately for geese, their natural world presents few objects that compete with eggs for their attention.

Tinbergen and his students decided to work out what it was about an object that signaled "eggness" to ground-nesting birds. They developed a classic two-choice technique in which they offered two dummy eggs to a goose or a gull, and then scored which egg the bird recovered first. This allowed them to compute the relative value of each morphological feature. They discovered that, up to a point, bigger is better; whatever impels the bird to initiate egg retrieval is stimulated more by an exaggerated cue. This "super-normal" effect is not limited to egg size: speckled eggs elicit a speedier reaction than plain ones (goose and gull eggs are normally lightly speckled), and larger speckles are more attractive than ordinary ones. The birds prefer ovoids to rectangular solids, and green eggs to the usual olive drab.

What we are looking at here is the response of recognition circuits to specific telltale cues. These innately recognized stimuli, which Tinbergen and Lorenz called *sign stimuli,* tell an animal when to produce a particular response and where to direct it. The sensory processing that isolates these cues need not be perfectly matched to the targets; all that is necessary is that the cues be sufficiently useful to trigger positive responses where appropriate while avoiding too many false positives. Where there is an ambiguity, the usual solution seems to have been to add an extra sign stimulus rather than to refine an existing one.

Sign stimuli organize behavior before experience can play a role, and in a surprising number of instances the result is so accurate that the animal needs no learning to improve either recognition or response. The programming that Tinbergen and Lorenz realized must underlie so much of animal behavior consists of sign stimuli, which are combined to produce a kind of signal, and motivation or drive—time of year or degree of hunger, for example—which sets the threshold of an *innate releasing mechanism* that can in turn trigger the appropriate motor program. If there are enough of the proper stimuli when the motivation is appropriate, the motor program is released. The performance of a motor program may then produce a new sign stimulus, and the next response in the sequence; a truant egg returned to the nest, for instance, releases the characteristically stereotyped hollowing behavior in which the bird's feet push out behind while the breast settles again over the full clutch. Where innate mechanisms play a major role in building, it is this kind of wired-in orchestration we will be looking for.

CONDITIONED LEARNING

Lorenz and Tinbergen also studied *conditioned learning,* though that is not what they called it. Conditioning was at that time the major preoccupation of behavioristic psychologists, who thought

Imprinting. These goslings memorized the sound and appearance of their parents thirty-six hours after birth. After three days of age, this identification cannot be altered or unlearned.

this ability to learn a connection between a stimulus (S) on the one hand and a response (R) on the other could explain everything that animals do, including egg rolling and nest building. Led by J. B. Watson, they believed that S→R conditioning operates in the same way in all animals, its power limited only by sensory abilities, the nature of the muscle and limb arrangement (which constrains responses—pigs can't fly), and the space in the brain available for learning and memory. They drew their inspiration from Ivan Pavlov's work on salivation in dogs. But behavioral conditioning as we understand it now was seen first in an apparently odd kind of learning called imprinting.

Many precocial animals, creatures that must move about on their own shortly after birth, need to memorize something distinctive about their parents as soon as possible to maximize their security in a dangerous habitat. Geese and deer are two familiar examples. But in a world filled with distracting and irrelevant stimuli, how is a newly hatched gosling to decide whether to attach itself to a parent, a pebble, a sibling, a cloud in the sky, or a marauding gull?

The essential prerequisite for conditioned learning, strangely enough, is that the animal must know something about the correct answer in advance. This knowledge serves to link preëxisting responses to novel stimuli, or novel responses to innately recognized cues; for this to happen, the animal must understand one or the other in advance, and recognize that learning through conditioning is the appropriate behavior to engage.

In *classical conditioning,* the kind originally discovered by Pavlov, Watson and the behaviorists called this partial answer an *unconditioned* (innate) *stimulus*; the unconditioned stimulus is the same as the sign stimulus of ethology. Goslings, for instance, are born ready to focus on such sign stimuli as movement away from the nest, a species-specific "exodus" call, and a certain waddling motion on the part of the target. In many ducks, a brightly colored patch on each wing called the speculum is also involved. All these cues together pretty well define a parent; given the limited range of cues that are available to one- or two-day-old chicks (the age at which parental imprinting occurs), nestlings in a natural setting have little chance of getting this wrong. In the lab, however, parentless chicks can become imprinted on a moving balloon, a shoebox, a toy, or a researcher as their motivation to attach themselves to something increases.

The goal of parental imprinting for the young animals is to recognize their parents as individuals; anything that is common to conspecifics in general is not useful. Once they are up and following something, what should they memorize about this particular individual? Should goslings concentrate on voice, face, wing patterns, or what? Getting this wrong would be fatal. What animals learn is called the *conditioning stimulus* by psychologists. We now know that animals have built-in adaptive biases that ensure they remember the correct cues. For pigeons being conditioned to recognize food, visual stimuli are key; this is reasonable because the seeds they eat are silent and the world around them is full of irrelevant noise.

But when the learning task involves avoiding danger, sounds become the innate focus of their attention. Here is another part of the answer that must be known in advance for conditioned learning to work.

The birds know even more before they begin learning. Pigeons do not blindly memorize everything visual about a feeding situation, or even about seeds. A built-in component of classical conditioning is a probability analyzer; this mechanism keeps track of the positive and negative correlations associated with each cue (color, shape, size, and so on for seeds), as well as the number of false alarms. The exact strength of conditioning can be predicted with striking accuracy through these values. In short, the animals know how to process the cues they must attend to, when they encounter the stimuli they innately know define a specific learning task.

Classical conditioning is innate and automatic; the cues that trigger it for each task, the cues that are memorized, the processing that sorts out cause from accident or negative stimuli, not to mention the storage and later use of this information, is an adaptive unit. This kind of learning is designed to allow an animal to move from crude sign stimuli to a detailed picture of the thing being learned, whether parent, food, building material, nest site, offspring, or predator. The intriguing thing about imprinting is how rote it is: it takes place only at a specific age, and it is usually irreversible.

Aside from learning to recognize specific individuals or objects, animals of many species need to learn novel behaviors, or how to modify innate programs to better match the realities of the task. Originally, behaviorists thought complex behavior was compounded from simple reflexes, as chains of crude prewired responses such as knee jerks and eye blinks assembled into sequences like walking or typing. B. F. Skinner proposed later that behavior in animals, although still reliant on these simple circuits, could take advantage of trial-and-error experimentation. This so-called *operant learning*, wired in but educable, would be the way a nest-building program could be adapted to circumstance, or improved upon if a basic plan were instinctive. Whether this optional cognitive add-on will be selected for in a species ought, in theory, to depend more on niche than on brain volume.

Like classical conditioning, we now know that operant learning comes with many biases specific to a species. If a task to be learned involves food, pigeons experiment only with their beaks; if they need to discover how to avoid danger, they try variations with their feet instead. (Pigeons outside of cages first run and then take wing.) This behavior makes perfect sense: pigeons never eat with their feet, and to peck instead of flee when danger is present would be suicidal. Rats, however, are likely to use their paws to explore food and to bite if threatened, and so they respond to experimental tasks differently. Operant conditioning, like classical conditioning, depends on already knowing part of the answer.

The two types of conditioning, then, are "learning to recognize" (classical) and "learning to do" (operant). When a chickadee learns first to recognize sunflower seeds, and then by trial and error how to open them, we see these two kinds of associative learning working seamlessly together. Each is widespread among animals, and we should expect that conditioning, with its attendant biases, will play a major role in the building behavior of many species.

Organ-pipe Wasps

Most examples of animal architecture occur out of sight. Birds' nests are usually well hidden from nest predators and birdwatchers alike. The soil is full of ants, yellow jackets, and voles, but only the comings and goings of these builders reveal their unseen communities. Many spiders take their webs down during the day. But on a shady wall, under the sheltering eaves of a house, there are likely to be the characteristic vertical tunnels of hardened mud built by organ-pipe wasps.

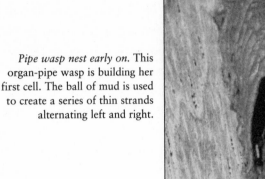

Pipe wasp nest early on. This organ-pipe wasp is building her first cell. The ball of mud is used to create a series of thin strands alternating left and right.

The construction of these pipes reminds us of how the several behavioral tools available to animals can work together to create a single kind of nest. The wasps are solitary, though on occasion two females may work side by side, sharing a common wall. Each pipe is about four inches long, half an inch across, with walls one-sixteenth of an inch thick. The inch-long female begins at the top with a ball of wet mud and applies it to the vertical surface, stretching it out into ropes braided into a herringbone pattern, creating a ∧ built in two steps: first the /, working from bottom to top, then the \, again beginning at the bottom and joining its mirror image at the top. This first line is roughly one-sixteenth of an inch wide. The next one is identical, except that the ends near the wall are narrower; it is attached to the outer, lower edge of the first strand, so that at its center it is separated from the wall.

After a few of these angled additions, the wasp has a genuine vertical tunnel that arches out half an inch from its support. In horizontal cross-section, about a quarter inch on each side extends

Pipe wasp nests. A female organ-pipe wasp has built five tubes side by side, each containing four or five compartments. About half of the larvae survive the risk of predators, parasites, and developmental failure.

directly out from the wall; these two side supports are joined by a perfect half circle of worked mud. As a result, the pupal wasps that will develop in single file along its length, three-eighths of an inch in diameter at maturity, will fit comfortably inside. The outside of the tube is relatively rough and dries to a concrete-like hardness; the wasp uses her mouthparts to smooth and polish the inside.

The builder interrupts her task every inch or so to collect food for her offspring. For this species the prey is spiders, which are paralyzed (so they will not spoil) and stuffed up into the cylinder. The tangle of spider legs holds each prey item in place. When the chamber is full the wasp lays an egg, builds an up-curved wall at the bottom, and then continues constructing the pipe. Small parasitic wasps watch for these structures; if the owner is away collecting mud or capturing spiders, the intruder will bore into the upper chamber and lay an egg (which later hatches and feeds on the wasp larva). In the process, the parasites build their own delicate and beautiful little mud tunnels inside.

In the usual course of things, that is—less than half the time, what with all that can go wrong—the egg of the organ-pipe wasp hatches into a larva that feeds on the cache of spiders and then pupates; the pupa is enclosed in a cocoon of silk that the larva weaves about itself just before it transforms into a pupa. The pupa metamorphoses over the next ten days into an adult wasp, which then chews its way out of the pipe to seek a mate and continue the cycle.

INSTINCT OR LEARNING?

We can readily identify many of the innate building blocks of behavior that are recruited to the building task. Adult solitary hunting wasps feed on nectar, for instance, but in provisioning their young with the protein they need to grow, they are often species-specific in their hunting: bee wolves take only honey bees, others rose-chafer beetles, tarantulas, cicadas, or tobacco hornworm caterpillars. In 1873, Douglas Spalding commented on the need for instinct in this behavior, writing that a hunting wasp "brings grubs—food that as a wasp she has never tasted—and deposits them over the egg, ready for the larva she will never see." Such specialization, although fascinating, is not surprising: the location of the neural ganglion that must be injected with poison to paralyze the prey differs from one species to another. A wasp that specializes on honey bees, for instance, inserts her sting accurately between two distinct plates on the underside of the bee's neck, immobilizing but not killing it.

Ingenious tests by J. H. Fabre, Niko Tinbergen, and others have shown that the wasp's recognition of its prey is innate, and that her attack and stinging behavior is a motor program. For some wasps, the sign stimuli involved in recognition are known. Odor is the cue that draws the bee wolf upwind. The sight of a small dark object releases an airborne pounce; tactile cues trigger stinging. Something analogous must account for the ability of organ-pipe wasps to recognize, subdue, and paralyze their spiders.

But learning is involved as well. Any hunting wasp must remember where her nest is located because she returns many times during construction and provisioning. And given the poor vision of insects (about 20/2,000 for this species), the wasps must use celestial and terrestrial landmarks on both large and highly local scales. She must integrate the angle of each leg of her hunting journey relative to the sun to guide her to the vicinity of home; since the sun tracks from east to west across the temperate-zone Northern Hemisphere sky, the wasp has to learn not only which direction it moves but also the rate of movement at different times of the day (which is substantially faster near noon) to compensate for the time that passes during a hunting trip. And although backtracking may get her at least near home, she locates the nest precisely through her memory of small landmarks.

The trips for mud for the nest require the same skills, as well as an ability to recall where the source of building material is found. Locating suitable mud in the first place is a problem: what is the sign stimulus for mud? A wasp's ability to select material of the right consistency gets better over time, suggesting some benefit from experience. And although the actual building of the pipe is so stereotyped that it must be based on a set of coördinated motor programs, later pipes show definite improvement over earlier ones.

A constant question in building behavior is how much the animal understands about the shape or function of what she builds. Experiments can help dissect this behavior by showing which sorts of alterations she will repair, and which she will ignore. The wasp, for instance, will investigate if the top of a pipe is removed; if the first cell is not yet complete, she will build a new chamber wall and continue. If the first chamber has been finished, however, she will ignore the problem and proceed as if the roof were still there. This points to a system that relies heavily on a preprogrammed building routine rather than any real idea of what the finished product should look like, or what function it serves.

The behavior of the organ-pipe wasp presents most of the questions about building and the neural processes that make it possible, but few of the answers. To tease out the cognitive mechanisms at work, our strategy will include looking at the ways in which the challenges of a species' niche are often a better predictor of mental abilities than phylogeny. We will focus on three comparisons, beginning with the uses of that most remarkable of biological products, silk. Then we will look at the most successful and diverse group of organisms on the earth—creatures that outnumber us 400 million to 1: the insects. Our third set of examples will be chosen from those familiar master builders, the birds. We will end with a look at some spectacular instances of apparent insight, and ask what the minds of animal architects have to say about our own species.

CHAPTER 2

Building with Silk

Silk is easily the most remarkable building material on the planet, and it has one source: arthropods. Humans have yet to produce anything like the strong but elastic substance of which cocoons, spiders' webs, and many other things are built or glued together. Silk is formulated in special glands to serve any of a variety of needs. It can be made literally stronger than steel, or, by a change in chemical proportions, more elastic than rubber; it can be astonishingly sticky or as slick as glass. And there is a lively secondary market for recycled silk in species as diverse as hummingbirds and humans.

Silk is a protein, as are the basic molecular components of muscle, hair, and tendons. Proteins are linear chains of amino acids that have the same central structures but differ in the atomic groups that decorate the sides. The core of a silk molecule is a pair of amino acids, alternating up and down:

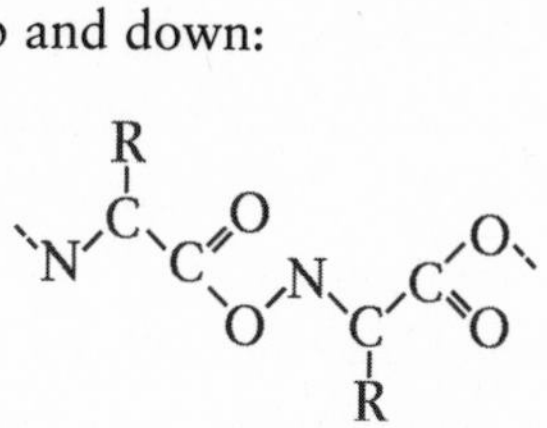

N is nitrogen, C is carbon, O is oxygen, and R is the side group that distinguishes one amino acid from another. A strand of silk has hundreds, thousands, or even millions of these chains running side by side in parallel, each of which is potentially billions of atoms long. The silk gland can produce an array of diameters as needed, but spiders extend this range enormously because their three gland types run from extremely fine to the heaviest gauge found in nature.

Silk's amazing properties depend on three things. First is its zigzag structure, which can be stretched to about three times its resting length. The bonds between the atoms resist (but do not prevent) the angular deformation required for stretching, but the bonds themselves are even stronger; thus the energy needed to break the chain is much greater than that required for bending the bonds. And all of the thousands of side-by-side chains must be stretched at the same time.

The second trick is that the zigzags in one chain can be aligned with those of its neighbors; when they are in alignment, the chains attach themselves to each other through weak electrostatic attraction—positive charge to negative. Although these bonds are individually weak (from one-tenth to one-twentieth the strength of the connections between the atoms in the chain), there are millions of them between every adjacent chain. Silk would be much weaker if the chains could each stretch independently; this electrostatic attachment is what makes silk up to twice as strong for its size and weight as structural steel.

The third bit of magic lies in the side groups (R). Some of these groups are very small, others are of moderate size, and some are huge. The range is from one atom to eighteen. The animal creating the silk controls the nature and distribution of its molecular configuration. If all the chains have small side groups, the silk is essentially crystalline; if the R-units are all large, the strand is almost liquid. Thus the creature can fabricate a piece of material with the diameter, strength, and elasticity suitable to the task, and the next strand—or the next part of the same one—can be completely different.

This chapter is about the many incredible uses natural selection has found for silk, and the behavioral strategies and cognitive equipment needed for exploiting this unique material—instinct, learning, and planning. When it comes to animal behavior, evolution regularly beggars imagination.

Threads, Processions, and Tents

Although the first things we usually think of as silk-based animal creations are webs and cocoons, a typical wooded yard has many more examples. In spring, for instance, when new leaves are soft and young, most trees have tens of thousands of caterpillars at work chewing away; the rain of frass, or caterpillar feces, from the canopy is a constant reminder. The caterpillars, being the food source of many species of nesting birds, in turn supply countless chicks with their primary nutrition.

But the depredations of birds are nothing to those of hungry wasps. The wasp menace increases steadily: because their colonies grow exponentially as the spring and summer wear on, they require a steady flow of caterpillars to feed their young. In search of this food, they scour the underside of every leaf at close range. Adding insult to injury, when caterpillars feed, the damaged leaf edges of some plant species emit odors that actively attract predatory wasps.

Caterpillars depend on silk for survival. Their simplest strategy is surprisingly effective: when disturbed, the insect larva drops from its leaf. Disturbances, such as a bird landing on a nearby twig or the airborne vibrations from a hovering wasp, intrude on the otherwise silent world of the caterpillar and innately signal danger. But the caterpillar does not just launch itself into the void; as part of its prewired motor program, it trails a silken dragline behind and hangs, supported by its own invisible bungee cord. In time, it either continues its journey to the ground or climbs back up to its leaf.

For caterpillars to make instant use of the drop-and-hang ploy, they must "know" how to keep themselves anchored to the leaf at all times. The dragline is usually affixed to the base of the current leaf, ready for instant use. The larva generally works its way from the tip back to the base, reattaching the line when it moves to a new leaf. Other species, many sightless, deploy silk guidelines lower on the tree that they play out on their commute to the feeding site; returning is a matter of following the thread back.

Some types of commuting caterpillars feed at night, when predators are in their nests. The larvae spend the day hiding in bark, under branches, or (for gypsy moths) on the ground. As darkness falls they climb back up and out using silk as a trail marker, often leaving a new strand on top of it. Later in the year, if you look closely when the light is just right, you can sometimes see a poorly organized crust of fine crisscrossing strands generally aligned vertically. But even a diurnal feeder benefits from the guideline tactic: Older leaves can be hard to digest, so that, like cows, their most efficient strategy is to alternate bouts of eating with periods for processing. Commuting away from the leaf for some digestive downtime is safer than waiting where the predators are looking.

Although none of this seems too cognitively challenging, we should ask whether these creatures have some sort of "picture" of their routes or the arrangements of branches they are feeding on. Some hint of this comes from processionary caterpillars. Processionaries use silk in two ways. These creatures live communally in a coöperatively constructed refuge—a waterproof and predator-resistant bag of silk suspended in a tree. Here a community can rest in safety and venture forth to feed (generally at night). The first caterpillar out leaves a silk guideline that the next one follows, laying its own on top. By the time the last creature leaves, there may be a visible silver highway leading away from the tent—a trail they will use later when returning.

At first, the foraging party heads upward: the leader climbs along the main branch and chooses a side twig, leaving a cable for the others to follow. Some species actually crawl head to tail as the column moves out. As the twigs become thinner, the caterpillars

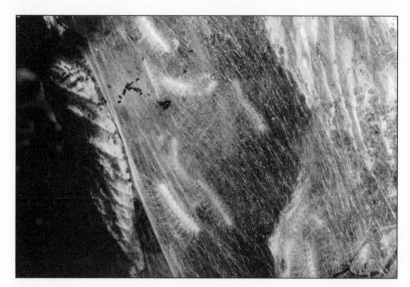

Tent caterpillar nest. Caterpillars build this silk structure coöperatively and retreat into it each day.

know they must spread themselves separately, one or two to a leaf. Returning to the nest is a matter of following the silk road back.

There is no formal organization to the group; the first to leave is trailed by the second. It used to be assumed that one caterpillar was a natural leader and that the others followed. In the late 1800s, the French naturalist J. H. Fabre put the hypothesis to the test by simply flicking the first one off the branch; the next one took up the leadership responsibilities without missing a step. When he diverted an individual in mid-column, the ones behind typically followed its silken cord; the colony then foraged in two groups for the evening.

Fabre's most famous experiment involved redirecting a line of caterpillars down a branch, across a yard, and up onto the rim of a flower pot. The first caterpillar laid a thread, seconded by the next, and so on, until all had circled the pot. If the column was long enough it formed a complete circle, head to tail, and Fabre had to direct the extra individuals elsewhere. Never reaching a branch, the column circled endlessly, each caterpillar adding another strand with every revolution. After several days broken only by periods of rest, the glistening silk highway was about a quarter of an inch thick.

Fabre, as was his wont, found this performance wholly mindless. Following a set of silk guidelines out and back, as most of the caterpillars in the colony do, may not seem intellectually demanding, but other elements are more intriguing. For instance, when he took pity on the group and directed them back to the ground, the caterpillars, though blind, set off straight for the home tree without needing to use the silk trail laid down days earlier. Was this luck? Did a chance breeze carry some special odor from home? Or were Fabre's caterpillars, like many birds, mammals, and adult insects, able to integrate the legs of the outward journey and formulate a route home?

Fabre also noticed that when the group has completely defoliated the branch holding the nest and returns home, some sort of memory tells them the next day that the old branch is no longer a suitable destination. The group sets off down rather than up until it reaches the trunk, and then marches out to leafier twigs. Again, what sort of processing does this require, and how could the same ability impact building behavior? The idea that insects might create a mental map and use it to formulate simple plans seems absurd at first. But, as we shall see, it's within the scope of some species. And if certain arthropods have this capacity, it's time to look again at the rest.

COCOONS

Most silk is produced by a pair of glands in the mouths or tails of insect larvae. The vast majority of living species are insects, and most of them have a larval phase. Generalizations in this huge and hugely diverse group are perilous: in some groups, such as the termites and grasshoppers, the animal that hatches from the egg is a miniature adult rather than a larva. And though most larvae lose their silk-producing organs when they pupate, a few adult insects retain them past the larval stage. Spiders, too, though arachnids rather than insects, are able to produce silk all their lives through glands at the tip of their abdomens or in their feet.

Much of the silk produced by insects is used in making cocoons; cocoons protect the insects in the vulnerable pupal stage, during which metamorphosis from the larval to the adult form occurs. Only some species make cocoons; for instance, nearly all moths do, but very few butterflies; all hymenopterans (bees, wasps, and ants) make cocoons, but no flies, and so on in a way that contradicts phylogenetic logic. Commercial silk is harvested from the cocoons of silk moths. Over the course of three days, the larvae spin about two miles of thin fiber into an impermeable ovoid designed to protect the pupae they will become from predators and from the elements. The creation of the cocoon is a miracle of patience and precision, combined with a deliberate randomness; it is the starting point from which a seemingly endless series of elaborations has evolved.

Although the supports are not often visible, a cocoon requires a framework. Some species (especially many bees and wasps) pupate in specially constructed cells or hollow tubes; honey bees in their comb and our organ-pipe wasps in their mud tubes are two examples. For them, the task of making a cocoon is fairly simple: the walls and the two ends of the cylinder provide a structural framework for the silk. Caterpillars and other cocoon-building insects must more often create their own scaffolding before beginning the cocoon.

To form the framework, many insect larvae begin by standing up on their convenient tail hooks and attaching a piece of silk to something near their heads. They bend to the side and search for a second point of contact, usually the ground, a leaf, or a branch the larva is standing on. They glue the strand down and return to a standing position, attach another fiber, and bend out in another direction. Eventually, the larva has an irregular and loose cone of fibers. It continues the project by running lines between different parts of the cone or from points on the interior of the cone to the surface underneath, creating the bounded cylinder that wasps and honey bees inherit, prefabricated.

In some species, this nearly invisible framework has no second support; it is attached only to the substrate. These larvae take advantage of the chemistry of silk to construct a stiff, almost crystalline enclosure that (if you can see it at all) looks like a Quonset hut made of chicken wire; they then attach the cocoon to this lattice. But for most species, choosing a building site with convenient points for attachment just within reach in nearly all directions is essential.

Then comes the cocoon itself, which is made of much heavier grade silk. The outer layer is woven between the thin support strands, glued down here and there:

As time goes on, however, and the first layer of the cocoon is laid down, the larva is able to attach its new work to an existing layer of cocoon silk. The animal makes S-shaped movements with its head while its body sweeps across the interior, creating an open-loop pattern. There are three basic patterns, and these differ only in the rate of sweep:

The first produces the most flexible material, with strands connected only to the last layer. The second is a chicken-wire arrangement, though with a much finer mesh and more pliable silk than is seen in the crystalline-hut framework variation. The structure is much tougher because the loops are glued to each other. Perhaps the most common strategy is the third: the loops overlap enough to be attached to one another at four places rather than two. This yields the stiffest cocoon of all.

Silkworm cocoon. After creating a scaffolding, the silkworm begins weaving its protective case around itself, using two miles of silk in the process. Once the cocoon is finished, it pupates and undergoes metamorphosis into the adult moth.

After a complete sweep, the larva moves by way of its tail hooks slightly along and about its long axis; the next line of loop stitching is thereby systematically displaced, but intersects the previous one. The resulting structure is very strong indeed.

For many insect larvae, the framework step is normal but optional; if the creature finds itself already mostly surrounded, it simply finishes over the gaps and then starts the cocoon. Commercial producers place silk-moth larvae ready to pupate into tubes of the right diameter; this technique ensures that most of the useless framework, which would only tangle the strands of the finished product, is never built. When the cocoon is completed, the pupa is killed by heating, and the silken cocoon is put into hot water. The glue that holds the strands to one another dissolves, and the cocoon begins to unravel. Workers separate out a loose strand from the outer layer and carefully reel the silk onto a spool. Since a single fiber is far too sheer to weave into clothes, threads from many cocoons are usually pulled at once; the film of dissolved glue that adheres to the strands binds them together.

The larvae's ability to adapt their building behavior to circumstance, even to human intervention, tells us something about how this behavior must be wired. When the drive to stop feeding and begin to build a cocoon is triggered, the larva must move to a suitable position, often some distance from where it has been feeding. The behavior is automatic, but it depends on variables: if the caterpillar can touch support surfaces right away, as with the artificial tubes provided by silk producers, it skips the framework step and begins spinning its cocoon immediately. More interesting is the question of how the creature judges whether the distribution of potential support points is satisfactory. For a human, working blind and groping experimentally with arms to map the space nearby, the cognitive demand would be substantial. But humans rarely encounter such a challenge; for the larvae, it is a life-or-death task faced nearly every generation. Selection must have worked to create specialized tactile mapping circuits to facilitate a job we would find daunting. But how could selection manage such a task?

Sensory Maps

This is a critical question, because sophisticated building behavior often depends on a series of internalized mental maps. The most basic of these neural representations is of the body itself. An animal usually "knows" where its tactile receptors are—knows in the sense that it accurately and appropriately responds to localized stimulation. This is quite different from the basic stimulus response (S→R) system that must have predated anything more complex.

In an S→R circuit, each sensory ending or group of endings—tactile receptors, say—is wired to a response circuit. At the cellular level, most neurons consist of relatively short information-collecting processes called dendrites, a cell body, and a long axon ending in synapses on the dendrites of other neurons or on muscle cells. The

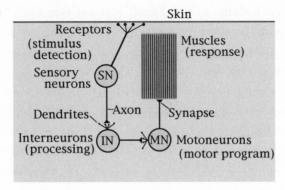

A simple neural circuit. The simplest circuits connect sensory organs to muscles. Different sensory neurons collect information in a variety of ways. In this example, the dendrites detect the stimulus and, if it is strong enough, cause the cell to send an action potential down the axon to synapses on the dendrites of interneurons. If enough sensory neurons fire at once, the interneurons relay the message to a motor-program circuit, where motoneurons orchestrate the response of muscles.

stimulus that the receptor is designed to react to causes electrical activity in the dendrites that spreads toward the cell body and creates a rapidly moving pulse called an action potential. The action potential sweeps along the axon and causes the synapses to discharge chemicals that stimulate the next cells in the circuit.

In most of these circuits, the inputs from numerous sensory receptors are integrated by one of the middlemen of the nervous system, an interneuron. (As with most neural wiring, the axons from developing neurons are drawn chemotactically to their target cells or regions, which may be either quite close or at some considerable distance. When axons grow, they tend to adhere to a previous axon going in the same direction; thus, over time, major nerve tracts—bundles of axons—link regions of the body.) The target interneuron activates the appropriate response circuit, where the motoneurons (nerve cells that synapse on muscles) orchestrate the response. For behaviorists, the sensory input is the Unconditioned Stimulus, and the motoneuron output is the Response. Learning at the neural level occurs by wiring up other sensory inputs to drive the Response; this means that other (initially suppressed) sensory inputs must feed into the response circuit. For ethologists, the only stimuli that the interneuron attends to are the sign stimuli; the drive is the response threshold of the interneuron. The cluster of interacting interneurons and motoneurons that produce the response are the basis of the *motor program*. Learning occurs in the way behaviorists imagine, except that only certain sensory inputs are available for each response, which accounts for the selectivity of learning; and drive can affect the process as well because it increases in intensity until the behavior is performed, leading to critical periods.

The reason a simple S→R system is unsuitable for anything but the smallest creature with a minimal behavioral repertoire is that each sensory receptor (or localized group of receptors) would need to be wired to a specific motor program—a response circuit that is directed at the particular receptor. As the animal grows longer, say, the number of receptors increases as the square of length; an animal an inch long requires ten thousand times as many neurons under this plan as one a hundredth of an inch long. Soon, either the brain would need to be larger than the body or the number and precision of the responses would have to be scaled back drastically.

The solution to this escalating problem of neural overhead is to invest instead in circuitry that interpolates; that is, motor program units that deal with stimuli over a wide range of locations by adjusting the response to target any intermediate location between (at the extreme) head and tail. But to do this, the sensory input needs to be organized in a logical, map-like way: head to tail, left to right, up and down. No one understands how such sensory maps evolved, but a likely scenario assumes that early creatures, like their modern counterparts, produced their receptors and other neurons sequentially as the animals developed additional segments during growth.

(The simplest creatures with body maps are all segmented; insects are the most obvious example, but vertebrates also develop from a series of internal segments laid down one at a time from head to tail.) The axons from later-developing receptors would have arrived at the brain or sensory ganglia later, more or less automatically forming a linear map; incidentally, they would also have preserved up/down and left/right relationships simply by growing the axons along the periphery of existing nerve tracts. Selection would have favored creatures that optimized this incidental mapping. We will distinguish between the primitive S→R system and the later under-the-skin maps of sensory input as Tier–0 and Tier–1 levels of neural organization.

Sensory maps are ubiquitous, a testament to the great value of sorting out stimuli in a logical manner. For instance, signals from adjacent facets of the compound eye are wired to adjacent neurons in the visual ganglia, the whole visual arrangement of up/down and left/right being preserved. Receptor activity from tactile stimuli almost inevitably projects onto ganglia in spatially organized arrays. Even stimuli that have to be processed abstractly to extract information about location, such as sound, seem to wind up organized into a map-like grid. There is no such thing as a compound ear, for example, with designated facets listening for noises from specific directions. Instead, differences in arrival time and loudness allow the brain to compute direction and then to plot it on a neural grid that gives rise to our sense that a given sound comes from a particular place in space.

With Tier–1 maps, the animal can react in a graded way, interpolating perhaps between various extremes. A certain degree of contraction or relaxation of the flexors, extensors, rotators, and other muscles in the shoulder, arm, and wrist, for instance, takes a paw automatically to the spot on the head that needs a scratch. A specific motor program for responding to stimulation of each tactile receptor on the scalp would fill the brain to overflowing, leaving no room for anything else. The evolution of an interpolation system is easiest to imagine if selection starts early, when there are few receptors to

manage (a low-resolution task), and just a few response motor programs to deal with the stimuli.

Only the evolution of sensory maps and response interpolation
makes the later development of large creatures with precise control
a possibility; that is, maps open the door not only to better maps
but also to the evolution of more complex behavior. Interpolation
also introduces strong selection on the arrangement and control of
muscles and joints. Imagine that, initially, two tactile receptors re-
port on stimuli impinging on the animal at two points on the body.
For each of these receptors, there are wired-in responses for dealing
with stimuli from these two locations. Sensonry interpolation allows a creature to react to a new receptor in between with a motor
interpolation—a response intermediate between the two hard-wired
reactions to the older receptors. Selection will strongly favor more
"logical" and linear response systems—neural circuits, muscle insertions, joint angles—that increase the accuracy of reactions. And
from this change arises the possibility of adding yet more receptors,
and perhaps even reducing the number of motor programs as precise interpolation makes many of them unnecessary.

The next step is the development of "personal-space" maps. Such
an ability would allow an animal to deal systematically with the
world just outside its body with the same precision and sense of location provided by tactile input from below the skin. These Tier–2
maps are likely to have arisen from duplications of existing Tier–1
tactile maps. Three major features stand out in brains: specific areas
dedicated to particular tasks, map-like sensory and response areas
organized in a map-like way, and map duplication. Primate brains,
for instance, have at minimum a dozen visual grids, each specialized
for a particular set of information-processing tasks. And each is a
mirror image of an adjacent map, arising because a genetic mutation
caused developmental genes to issue two sets of building orders instead of one. Indeed, much of evolution depends on duplications at
the level of single genes, followed by selection for independent spe-

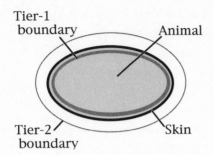

Tier-1 boundary   Animal

Tier-2 boundary   Skin

The first two cognitive tiers. Nearly all animals have a Tier–1 sensory map— an organized projection of the sensory receptors on and under the skin. A creature generates a Tier–2 map by probing and touching the part of the outside world within its reach.

cialization of the superfluous gene product. In brains we see the same strategy at work on a grander scale.

The most likely scenario for the evolution of outside-the-body maps is from a duplicate tactile map generated by a typical developmental mutation. This superfluous map would still have drawn axonal endings along nerve tracts from receptors, but (with two targets releasing the chemical stimuli that lead the growing axons) the sensory cells formed synapses on both areas. As is typical of duplications, the second map would not have been needed for passively representing where an incoming stimulus was being felt under the skin. Through selection, it would instead have begun plotting where an external surface or object could be found through active probing: a very small change in wiring, but one that opens up a huge range of behavioral possibilities. Building, for instance, depends enormously on probing to determine where potential supports are located, both relative to the animal and in relation to one another. Circuitry to make use of this information while fabricating simple structures would have to evolve, not only making a hard task progressively easier but also providing the opportunity for shortcuts and innovation. Such cognitive shortcuts complicate our efforts to interpret much of the building behavior we encounter in nature: the superficially difficult may be, at the neural level, trivial, whereas something that looks easy might require total concentration. And this complexity, as well as the opportunities that

Mapping in Animals

Tier 0:	No spatial representation; independent S→R wiring for stimuli.
Tier 1:	Internal map: spatial representation of stimuli impinging on body; typically tactile.
Tier 2:	Surround map: spatial representation of objects and surface immediately around animal (within one body length, typically mapped by touching); generally tactile.
Tier 3:	Local-area map: spatial representation of local objects not within one body length, allowing local navigation through interpolation and pattern matching; typically visual, tactile, olfactory, or auditory.
Tier 4:	Cognitive map: spatial representation of the relative position of widely spaced objects or other landmarks, allowing home-range or nest-interior navigation based on a cognitive map; typically visual or tactile.

appear as an unintended consequence, increase exponentially, as we will see when a third tier of larger-scale maps appears.

BALLOONS AND CASES

Though the egg cases of spiders look superficially like cocoons, they are constructed quite differently. Instead of being built from the inside, they are woven from the outside, around a cluster of eggs. The spider may carry her egg case with her to protect it, or she may camouflage it discreetly with sand, vegetation, or debris.

Case building in spiders has been modified from their habit of wrapping prey, a behavior in which time is of the essence. Spiders have no chewing mouthparts; they must paralyze or kill their prey, inject it with digestive enzymes, and then suck out the resulting soup. Most spiders wrap their prey to some extent to hold the victim still while injecting it.

Orb weaving spiders are champion prey wrappers. When it senses an insect struggling in its web, the spider rushes to it, attaches silk, and begins to roll it. The spider typically rolls the prey item 360 degrees, rotates it a quarter of a turn, gives it another full-circle roll, rotated, and so on. Spiders, therefore, work in rapid straight stitches; there is no time for the painstaking S-shaped weaving seen in cocoons. The victim is its own scaffolding.

A nice elaboration of this is seen in some wolf spiders. The male will capture a fly, wrap it briefly, inject digestive venom, and then finish the wrapping job to create a beautiful ovoid, entirely free of the lumps seen when an insect is wrapped in a web. This is, in a real sense, gift wrapping: The male uses the fly to entice females to mate—a free meal in exchange for sex.

Some balloon flies use the same tactic. A male captures prey, wraps it in an envelope too symmetrical to be either accidental or efficient, and then uses it to seduce a female. The balloons, like much commercial packaging, are out of proportion to their contents, and they may fool the female into thinking the gift is larger and more nutritious than it really is. And then there are cheater males that suck the fly dry before wrapping it. More primitive members of the group make a simple jagged parcel, or do not wrap their gift at all.

One group of these flies has taken things a step further: the males construct completely empty balloons and then use them for courting. This behavior probably began as simple cheating by males too small, clumsy, or unlucky to find prey. Though these silken artifacts are impressive, the only suggestion of a cognitive component is in the way males decide whether and how to try to fool females—and the tactics females employ to avoid being exploited (such as hefting the package, or toying with the male while trying to get her mouthparts through the silken envelope to sample the contents).

Seine Nets

Moth larvae weave safe and weather-tight protective shells to pupate in; spiders use silk to protect their eggs and to catch prey. Their silk is strong and delicate, and the weaving, particularly among spiders, is often intricately patterned. But caddisfly larvae, which live underwater in running streams, take the ability to create a precise mesh to its limits. Some species use silk as a strainer to filter edible material from flowing water.

Most species of caddisfly larvae build open cocoons into which they can retreat. Grasping the cocoon with its tail hooks, the larva pulls its home around it by its stubby front legs. The caddisfly's trailer home tapers toward the rear, which is open; as water flows through, the larva strains it for food, and the current carries away the wastes. As the larva grows it adds new silk at the opening, decorates it, and often uses its rear hooks to break off a now-unnecessary bit of tube. Many species embed small stones or shells in the silk to provide additional protection or camouflage, and some of their tubes can be quite beautiful. The decorations are specific to each species, and can be used to identify the nondescript larvae.

It's not much of a stretch to imagine how the typical caddisfly strategy could be programmed. But one species that lives in sluggish streams builds quite another sort of structure—a large funnel-like device, its entrance two to three inches across. The silk funnel is woven in a rectangular mesh; water passes through, but larger things, such as food and inorganic matter, are blocked and carried to the bottom. The larva waits on the stream bed or a large plant stem below and picks through the takings.

Although these free-standing nets are spectacular, the really clever filtering is done by members of another branch of the family. These larvae expand their tubes by attaching them to a larger chamber, which they construct from sand-encrusted silk. The chamber has two chimneys, one very tall and pointing upstream. The water enters in part because of the way the opening is oriented, but the structure also takes advantage of the Bernoulli effect: liquid passing over the short chimney draws water out of the chamber.

The larva builds the structure so that the wider head end of its tunnel opens into the upstream end of the chamber. It constructs a

Seine net. This species of caddisfly larva uses silk to build a large porous net facing into the flow of water. Material captured by the net sinks to the bottom, where the larva sorts through it.

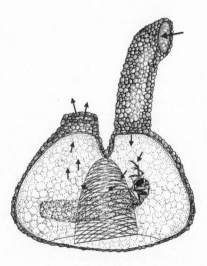

Caddisfly filter chamber. This complex structure is fabricated from silk, to which sand is attached on the outside. Water flows in the tall chimney, through the silk-mesh filter, and out the short chimney. The larva lives in a side tunnel that connects the main chamber upstream and downstream of the filter, and reaches out of the tunnel to sort through and eat anything trapped by the mesh.

Water flow

rectangular mesh inside across the long axis of its armored and camouflaged trap. The weaving is a greatly exaggerated and modified version of the S-shaped stitching of insect larvae in general. Standing on its tail hooks, the larva bends first one way, attaching the silk to the wall; it rises, using the wall itself to define the farthest extent of the pattern, attaches the silk again, bends back and down the other way, attaches at the next point, traces up the wall, and so on. Each attachment is located just past the previous one, and when the silk crosses the most recent thread worked from the other side, it is glued. At a distance, the pattern of larval movement looks like a figure–8, but up close the result is a fine rectangular grid.

Because of the nature of the shape to be woven, every cycle is different from the last, but the result could hardly be more regular. The extraordinary complexity of the task, and its one-time nature, point to intricate programming and a careful attention to real-time feedback during construction. And yet there is a flexibility, an ability to respond to unlikely contingencies that convinced Don Griffin, a pioneer in the study of animal thinking, that some degree of planning and spatial sense must be involved, something beyond the limited second-tier world of what the animal can touch at the moment. Does the caddisfly have the kind of third-tier map that would give it a useful larger-scale picture of its dwelling—the space just beyond its immediate (second-tier) grasp?

For experimental purposes, the best test of Griffin's intuition is to interrupt the building process, or to create some controlled damage in a finished part of the structure that defeats the purpose of the artifact. Most animals are baffled by such alterations, but not these insects. Caddisfly constructions are so small and difficult to start and maintain in the lab that such experiments have rarely been possible. But one species that builds a sand-covered dwelling was the focus of a revealing study in 1933. The house is fairly simple: atop the basic tube the larva constructs a gently curved roof that extends about one tube diameter to the side (rather like wings) and about two-thirds of a tube length to the front (an "awning").

If the rear part of the tube and wing is removed, the larva repairs the damage. What is deeply surprising is that in this experiment different larvae chose to repair their constructions in different ways. One built a new home from scratch; another extended the *front* to make up for the length lost at the rear; a third lengthened both ends; a fourth built a new front end on the back; a fifth did the same, but then modified the old front end to get rid of the awning; and so on.

The variation stands in striking contrast to the stereotyped egg rolling of ground-nesting birds. The flexibility suggests that the larva has, in addition to innate recognition and motor programs for accomplishing specific preordained tasks, some map-like idea of the finished product. This picture, whatever form it takes, is what allows the animal to find alternative solutions to the same problem: what we call a goal-oriented response. Because the animal seems to understand the goal of the behavior, it is freed from a blind compulsive reliance on task-directed responses. If birds were this smart they would try pushing eggs occasionally, and they might ignore the grapefruit.

The use of third-tier maps need not be very sophisticated. A drive for simple searching or experimentation continued until a pattern is matched is usually all that is needed. Whether finding the correct place for viewing a memorized set of landmarks, or producing an artifact that matches an innate or learned picture, the emphasis is on chance exploration rather than understanding. This is analogous to the progression from the passive experience of first-tier maps to the active self-generated touching needed for second-tier mapping.

There is good reason for caddisflies to have evolved third-tier mapping. In the natural course of things, some of these larvae modify their dens after they appear to be finished, repair damage, or renovate an existing den whose owner has either died or discarded it. Thus the experimental damage does not present wholly unnatural problems, but rather extreme instances of potentially realistic challenges.

This is a central point in analyzing building behavior: only if nature presents complex challenges is an animal likely to evolve the ability to deal with them. But whether its solution will be a series of wired-in task-specific backup programs or a goal-oriented response is the critical question in understanding how building behavior is organized. Any particular group could, depending on the long-term contingencies of its niche, have evolved either a rigid but reliable set of innate responses or a less foolproof but more flexible response— even if it is based on nothing more than choosing among and orchestrating the order of a set of prewired motor programs. We will be looking carefully for evidence of mental pictures, maps, and goal-oriented behavior as we examine animal architecture and the challenges animals must surmount.

COGNITIVE MAPS

Other creatures that use silk to hunt seem to require a fairly detailed memory of where the hunter and its prey are located, something more than a map of personal space or the larger region that encompasses a small nest. How likely is this kind of fourth-tier complexity in a six- or eight-legged animal whose brain is smaller than the head of a pin?

Psychologists and cognitive ethologists debate what criteria we should use when discussing "planning" and "thought." One threshold for this complex sort of mental activity is the ability to string together two separate and irrelevant experiences into a behavior sequence appropriate for quite another problem. Thus, when a chimpanzee drags a crate under a suspended banana, fetches a stick, climbs up, and knocks the banana down (as Wolfgang Köhler first observed in 1917), it seems likely that the problem has been solved in the animal's mind first. After a bit of cognitive (as well as physical) trial and error, the chimp executed its plan. Edward Tolman later called this plan a "cognitive map."

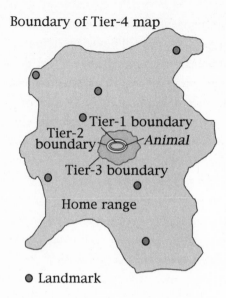

Boundary of Tier-4 map

Tier-1 boundary
Tier-2 boundary
Animal
Tier-3 boundary
Home range

● Landmark

The first four cognitive tiers. Outside the Tier–2 region (the region an animal can map through probing) is the Tier–3 locale area, generally the interior of a large nest or the region immediately surrounding it within which the creature must navigate locally back to the entrance. Some species map the home range into a wider-area Tier–4 representation.

Tolman did not mean a literal map; he envisioned rather a mental flow chart in which behaviors are mapped in sequences that yield results. The ability to plan can involve real-world spatial mapping skills, though. Close your eyes and imagine what lies twenty feet to the north, or a hundred feet directly behind you; or plan a novel and painfully indirect route to some familiar goal. Your ability to do this requires a mental map, and the ability to manipulate it and form a plan. In mammals, this activity has been traced to a small area of the hippocampus, a structure deep in the brain. The map involved can encompass the entire home range of the animal. Our intuition is that this fourth-tier level began as a scaled-up version of the third-tier maps used in more local contexts. When other animals form novel routes to get them to relatively distant goals, we must suspect they have this fourth-tier ability, too.

Few invertebrates turn out to be convenient for map tests. The first experimental demonstration of a cognitive map was in that easygoing behavioral workhorse, the honey bee, and even that result was hotly disputed for a decade by scientists rather too protective of the alleged cognitive uniqueness of warm-blooded vertebrates. Hunting spiders are better suited to this sort of manipulation; they are willing to run a maze for prey in full view of researchers. These salticid spiders (from the Latin *saltare,* to jump, because they typically pounce on their prey) are unusual: they have the best eyes among invertebrates by a large margin. They can see about as well as lizards because, unlike insects, they have a lens and a retina rather than a compound eye. In nature, they venture from their silken refuges to track down prey in a complex world of branches and leaves.

When a hunting spider sees a victim, it sets out in pursuit. In the three-dimensional world of a bush, there may be no direct route to the prey. This quarter-to-half-inch hunter will eventually have to return to its own home. Tests in the lab have presented the spiders with prey and artificial branches so arranged that the predator must actually walk away from its goal. The spiders each scan the maze, then about 90 percent of them set off along the reverse-route detour. Periodically along the way, each spider stops and rescans the maze. It is not clear whether it has forgotten the entire route, has made detailed plans for the first leg of the journey only, or is checking to see whether the prey has moved. But if the spider had not roughly comprehended the situation at the outset and come up with a tentative route, how could it have known it should set out along a path away from its prey?

In any event, the salticid has formulated a plan of some sort. It would be unwise and ungenerous not to think that other spiders and insects might have evolved similar talents where their niche requires them. The processionary caterpillars and caddisfly larvae, for instance, despite our deep and automatic prejudice against worm-like body plans, certainly merit consideration. Let's look then at some other silk users that could benefit from an ability to remember landmarks and plan routes.

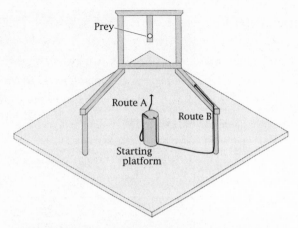

Hunting spider detour. The salticid is placed on the starting platform with a view of the prey at the end of the elevated maze. Most species would adopt Route A, moving directly toward the prey item, only to find it far out of reach. The majority of salticids instead move along a reverse-route detour away from the prey to the maze, and then find their way to the food.

REFUGES AND LAIRS

Salticids are only one of the many spider species that capture prey without webs, but they use silk in other ways. Most construct a lair, which provides protection and a suitable place for short-range forays in search of prey. Tarantulas construct elegant hunting lodges, excavating a rising tunnel into a bank and lining it with silk. The silk lining prevents the earth from collapsing or cracks from forming. Its strength is easily demonstrated; we can dig out the bank and separate the tunnel from the soil. The cylinder of silk is long and broad, and perfectly stiff.

Tarantulas, salticids, and other spiders that hunt from a lair have to know enough to get home after an excursion. Vertebrates use a couple of well understood ways to manage the return journey. One method we've already encountered: Using dead reckoning, individuals keep

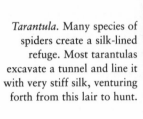

Tarantula. Many species of spiders create a silk-lined refuge. Most tarantulas excavate a tunnel and line it with very stiff silk, venturing forth from this lair to hunt.

track of the distance and direction of each outward leg; the animal integrates the legs at the end, derives a homeward distance and direction, and sets off. This method works best in two dimensions; in a more complex habitat, vertical travel must also be taken into account, and a simple straight-line path back to the lair may not exist. Moreover, to know the direction of each leg, the animal must have a compass. The directional guide involved is generally the sun, which creates a problem for creatures hunting at dusk or under dense vegetation.

Other navigating animals manage their return through a memory of the relative location of landmarks. (And many species—honey bees, for instance—can do both, choosing the best strategy based on the information currently available.) If you were kidnapped and set free at a familiar nearby food market, you would have relatively little trouble getting home. You would know where you were because you recognized the local landmarks; then you might either use a memorized route home or use your cognitive map to imagine one. But except for hunting spiders, arachnid vision is poor. A non-salticid spider requires large, unambiguous cues. And even with ordinary eyes, how distinct is one part of a bush compared to another? Although no one knows for sure which (if either) method is involved (and perhaps it may be a blend of both), it is clear that some sophisticated neural processing is required, and quite likely some cognitive ability.

One kind of lair-based hunter needs no obvious spatial ability. Trapdoor spiders excavate a tunnel around seven inches long and half an inch wide into a gently sloping bank. The spider lines the tunnel with silk and then fashions a lid compounded of silk and soil that has such an exact fit that water never gets in. The camouflage and precise fit make the chamber door almost impossible to see.

These creatures hunt strictly by ambush. Once night has fallen, they push the lid open slightly and extend their front four legs, back feet anchored firmly in the soil, or even on vegetation extending radially. Sensitive organs in the legs detect tiny vibrations in the ground created by the movement of passing insects; the spider compares the strength of the vibrations at each leg and so triangulates the prey's location. When the victim is close enough, the spider springs, never letting go of the lid with its back two legs. There is no way it can get lost, and perhaps second-tier mapping could account for everything.

Other species of lair-based spiders have evolved more elaborate ways of extending the range of their sensitivity to passing prey. A corolla spider from the desert of Namibia excavates a burrow and lines it with a tough layer of silk. The tunnels are about four inches deep and a quarter of an inch across. What makes these spiders special is that they arrange relatively large stones around the opening, attaching them firmly to the tunnel with silk.

But these are not just any stones. For one thing, they are nearly all quartz, which the spider chooses from the mixture of sand and gravel that surrounds the nest. The stones tend to be asymmetrical, and the spider arranges them with the tapered ends pointing toward the burrow opening. And although the range of sizes selected is limited (from two to three times the builder's own weight), an individual spider is flexible about the number of stones used; most lairs are surrounded by seven or eight pebbles, but the number can vary from six to twelve. If the stones are removed, the spider may well rebuild with a different number of building blocks and use components of different average size. The total combined weight of this new corolla, however, is generally within 5 percent of its previous weight.

The stones are no mere decorations; they are tools. Each serves as a kind of sounding board for ground-borne vibrations. The tapering arrangement increases the surface area available for collection. The hunter waits with at least four of its feet on individual stones, localizing the minute vibrations generated by walking prey. The stone circle efficiently increases the area of ground contact beyond that of the cross-section of a spider foot.

The flexibility of the spider's construction behavior, which includes the ability to make appropriate choices from an unpredictable array of materials and to create an effective signal-collecting device from quite different building blocks, appears to be more than ordinary programming. Cognitive ethologists might classify this as goal-oriented behavior: it fits our definition in that the animal seems to have an array of innate and learned motor programs that are available for use, as well as a strong drive to complete a task; but it needs to work out the details of the project according to the many contingencies of time and place. Of course, the main evidence for this remarkable mix of consistent precision and variability is sometimes negative: The researcher simply cannot imagine how the behavior might be programmed. Our own cognitive shortcomings, however, are probably not the best guide in such cases.

"IRREGULAR" WEBS

When we think of spiders, the famous orb web usually comes to mind. But in the early morning, before the dew has evaporated, bushes and tall grasses sparkle with horizontal circles of silk. These

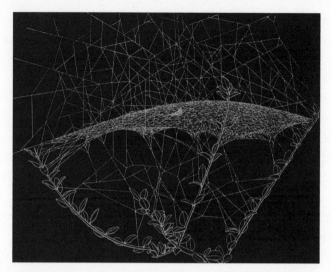

Hammock web. Some spiders build a loose hammock supported by threads both above and below. Prey knocked down by the overhead threads falls into the hammock, and is captured as it tries to regain its footing.

hammock webs, up to five inches across, are built by so-called money spiders. Their structure is marvelous: a sheet of finely woven but irregular mesh is held in place by a forest of scaffolding above and below. In some species, the main sheet is shaped like a shallow bowl; in others, it takes the form of a gentle dome.

To construct its web the spider begins with the peripheral scaffolding threads, first in the horizontal plane to define the outer edges of the hammock, and then vertical ones to give it support. Then the spider simply attaches a bit of silk to one point on the circumference, walks to another point, pulls the thread taut, and glues it. Patient repetition creates the mesh. Support strands, called stopping threads, on the interior of the sheet keep it from sagging. No two webs are the same, based as they are on the spider's size and the location of nearby supports.

None of the strands in any part of this structure is sticky. How, then, does the spider contrive to capture anything, especially considering that most of the encounters occur at night? When clumsy flies blunder into the network of stopping threads, or a katydid strikes one while attempting to jump from one branch to another, the insect falls onto the hammock. Its struggles to regain its footing alert the spider, who is waiting on the underside of the web. She rushes to the spot and bites the insect through the sheet.

There is great variation in prey size and behavior, and consequently a range of techniques necessary for capture from which the spider must choose. For instance, if the prey is large, it must be bitten several times. Its struggling may tear a hole in the web, so the spider instantly strengthens the weak spots and continues injecting the insect in different places.

But money spiders are not compulsive web strengtheners. If the prey proves too fleet, the spider may tear a porthole in the hammock and chase the insect on the surface rather than from below. If the prey freezes, the spider will wait, and then with an apparent loss of patience begin plucking on strands; this often starts the insect moving again, allowing the spider to localize it. If its prey has managed to climb a support thread or catch hold of one before falling into the hammock, the hunter may begin plucking near the base of the superstructure in an effort to dislodge it and so cause it to fall onto the web.

Do the many strategies among which the spider must select suggest cognitive ability? Does its seeming ability to picture where the prey is located and to negotiate a path directly there indicate some sort of crude map? And as it constructs its web, how much does the spider need to keep in mind about the structure of the space within which it is building, and what elements it has already put in place? Has its personal-space map been scaled out to encompass the web as a whole? These behaviors imply that more than mere stimulus and response is at work.

Some sort of third- or fourth-tier spatial sense would be useful to these spiders because, like other hunters, they have to keep track of both prey and home. Watching one pick its way purposefully across

Orb web. The owner of the web waits at the center listening with its legs. Gaps in the spiral catching threads show where prey has been trapped, rolled up, and cut out.

its irregular web, straight to an insect feigning death, it is difficult to believe that the choice of path is merely an ongoing response to immediate cues. But experimental work on this behavior has been confined largely to the distantly related orb weaving spiders.

ORB WEAVING

The classic orb web is the most regular structure made by spiders, and hence the most studied. It is a less complex achievement than the hammock web we just looked at; in general, the "irregular" web-building spiders evolved from the orb weavers, and they may represent a step forward cognitively. The very regularity of an orb web hints at an underlying simplicity. At a glance we can see that there is a radial framework and a spiral of silk. The web is built in a plane, and its weaver sits either in the middle with each foot on a different radial element or out of sight at the end of a signal thread attached to the center.

Detection and localization of the prey are well understood. The vibrations induced by struggling insects are transmitted along the slack and sticky spiral of catching threads to the taut radii, and thence to the center. The spider compares the strength of vibrations between radii to interpolate the angle of the prey from the center. In theory, the spider should simply rush out along the closest radius until it encounters the insect. However, when vibrations are induced experimentally, the spider nevertheless stops at about the right distance out from the center of the web. How is this possible?

We localize sound in two ways. For higher frequencies, our nervous system compares sound intensity between our two ears. If the loudness is the same, our brain perceives the source to be directly ahead (or behind); otherwise the difference in intensity specifies the angle to the right or left. The spider does the same thing with vibrations sensed through its web. Because it has eight vibration-sensitive legs instead of two ears, the question of in front or behind does not arise.

We place lower frequencies in quite a different way. Our nervous system judges the arrival time of a sound at each ear, and then measures the interval. If there is no delay, the source is straight ahead or in back; for each other direction, there is a specific time difference. A little bit of neural trigonometry supplies the answer. It is this time difference, measured among different legs, that allows spiders to triangulate the distance, which is something we cannot do.

Do these creatures have a mental picture of where their struggling prey is located? In theory, we might view the processing as a kind of first- or second-tier map, with the web as a functional extension of the body. But experiments suggest there may be more involved. In the lab, researchers vibrate multiple points on the web at once, and then they remove the stimuli. An orb-weaving spider can scoot out to one target, deal with it if there is an insect there, return to the center, and then, with only memory to guide it, launch off for the second one, return, and then pursue a third. At the very least, the predator can store multiple coördinates.

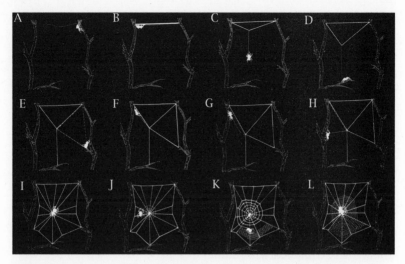

Orb web construction. The spider begins by releasing a very thin strand of silk with a drop of glue at the end. If this strand encounters a support and sticks, the spider hurries across, leaving behind a heavier gauge silk line. The spider walks back along this line, which stretches under her weight, and drops down in search of another support, playing out a new line of silk as she goes. She now has three of the radii of the eventual web. She continues to work on the radial lines, then adds a broadly spaced spiral beginning at the center. The spiral acts as a scaffolding for building the tightly spaced sticky catching threads. During this process, the spider removes and ingests the spiral.

Web construction behavior, along with the fact that webs are typically built in the dark, also suggests that a spatial representation may exist inside the spider. The spider climbs to a high point and releases a light, thin strand that has a sticky end. With even the faintest breeze, it wafts out to an unpredictable anchorage. When the spider finds tension on the line, she walks across it, leaving a much heavier line behind. Pulling this second bridge line taut, she has the top support in place. One axis of the emerging web is now defined.

She crosses the bridge to her starting point, leaving a loose medium-weight line behind. After securing it, she walks back to the middle of this strand, which sags both from her weight and the extra length she has spun. Now she lets herself down on a new line in the darkness below, until she reaches a new support. What with the inevitable pendulum-like swinging and perhaps a breeze, this bottom support is not usually directly below the line of the top crosspiece. Once she glues this bit of framework in place, the plane of the web is defined.

Then she returns to the center, glues a new piece of silk in place, walks out one of the diagonals, fastens a new line at the end, and then walks or drops down. This strand, once glued, becomes part of the outer framework. Now she can attach a new piece of silk, walk back up the side and then into the center, pull the line taut, and have another radial line. And so it goes until the entire web is framed and equipped with radii.

Once the scaffolding is up, the spider glues down a widely spaced spiral of threads, working from the center out. With this walkway in place, she lays a tight spiral of sticky catching threads from the outside in, gathering the walkway silk as she goes; this silk is to be eaten and recycled later. Once this circular mesh is finished, most species cut a hole in the center and reinforce its edges; now the spider has access to both sides from the middle. Finally, some species add a very obvious piece of white silk zigzagging from the center down, making the web obvious to birds that might otherwise fly into it.

How much of a mental picture does the spider need to accomplish this classic bit of animal architecture? We've seen evidence to suggest that some spiders can form and use third- or fourth-tier spatial maps. Would such a map-like representation make the building of orb webs any easier? Presumably any picture of the relative position of the supports for the framework could be analogous to the mental images we form when engaged in the awkward business of exploring a dark space.

In general, unless they have experience with blindness, humans are not very good at this task. The problem seems to be that our overeager brains seek to impose visual patterns on our tactile experiences, often with absurd results. But to a person with no visual expectations, the task is much easier; individuals who are not born blind but later lose their sight can draw reasonably accurate third- and fourth-tier maps. There is no reason to suppose that animals for whom the ability would be of great selective advantage would not, within the limits of their sensory equipment and brain mass, be able to do something similar.

If this were so, then we should expect the orb weaver to have at her disposal all the innate releasers and motor programs necessary, and arranged in a default order. But every set of initial anchor points is different; the number of radii is contingent on opportunity; the beginning of the sticky spiral depends on where the longest several radii turn out to be. In short, each web is a custom production. And orb weavers can repair unlikely damage (as when researchers snip one of the radii), and confound all expectations by constructing normal—indeed, rather better than normal—webs in the absence of gravity: aboard the space shuttle, spiders that ordinarily depend on gravity to fix the third (bottom) radius deal with this remote contingency with positive nonchalance. How, if the behavior is entirely based on task-directed programming, would this be possible? With even a third-tier map, the problem has a completely different solution.

TOOL USE?

We will look closely at tool use in later chapters, especially where animals build the tools themselves as opposed to seeking out and using natural objects. Tool use intuitively suggests a comprehension of cause-and-effect logic, of purpose and goal. In some sense the seine nets of caddisflies and the mud tunnels of organ-pipe wasps are tools, built by individuals to capture prey or protect young. And yet, we would hesitate to call our own houses tools. We are more comfortable with the concept (and the idea that understanding might be required for their use) when the tool is small and portable, such as a hammer or a drill.

Among the arachnids there is one group, the ogre-faced spiders, whose behavior involves seemingly unambiguous tool use. The net-casting spider, *Deinopis subrufa,* is a typical example. These highly evolved creatures have huge eyes (they are nocturnal hunters), are as much as three inches across, and build a small one-inch net of very springy silk. The spider begins by creating a basically rectangular three-inch frame of conventional thin silk attached to adjacent vegetation; the behavior is identical to that of a typical orb weaver, the group from which the net-casters evolved. The spot chosen is a likely commuter route for terrestrial prey.

With this scaffolding in place, it creates its one-inch net from numerous parallel strands. These bundles of silk are combed with special structures on the hind leg to create a loose tangle that makes adhesive unnecessary: any prey in the net becomes hopelessly enmeshed the moment it begins to struggle. The spider then cuts away most of the scaffolding, holding the four corners of the net at the ends of her front legs. When a victim approaches, the spider casts the net onto it and then kills it.

Net-casters combine the proactive hunting of salticids with the elegant building of orb weavers. Their web is a somewhat portable weapon that requires none of the passive wait-and-see strategy of most spiders. But what degree of comprehension is necessary in fabricating a species-specific tool? Does the web-caster understand anything about its building or tossing behavior? What sort of evidence would we find convincing one way or the other?

Invertebrates may have little or no imagination beyond stalking plans, spatial maps, and an ability to choose among alternatives in a generally sensible way. In the end, it is more parsimonious to ac-

Possible tool use by a spider. The net-casting spider, *Deinopis subrufa*, builds an elaborate trap net that it holds with its four front feet while waiting for prey to pass near. When a victim walks within range, it snaps the net onto the prey, or even tosses

count for many examples of spider and insect construction by inferring a basic cognitive mapping capacity, combined with an adequate ability to learn and remember, rather than resorting to explanations that require elaborate programming capable of anticipating every conceivable contingency. But even this is a huge upgrade in the cognitive status formerly awarded to arthropods.

The intellectual abilities of the silk-wielding spiders and insects we've discussed seem highly suited to the needs of their niche; indeed for some the construction behavior has created a major niche shift, possibly even a broadening. And the niche shifts, in their turn, will sometimes have selected for more flexible mental processing. The small size and complex habits, and the difficulty of rearing most spiders and caddisfly larvae in the lab, rob us of the chance to pose calculated and controlled problems to these intriguing animals. It also makes it hard to ask the more interesting question of whether selection has, as seems likely, left closely related but environmentally unchallenged species with less in the way of cognitive equipment. The differences within caddisfly species, for instance, or among spiders, or between social and nonsocial caterpillars hint at such a pattern. We need to take advantage of the relative convenience of insects to explore more systematically the limits and organization of innately directed building, and the more intellectual add-ons that have evolved when needed within subgroups of related but ecologically different species.

CHAPTER 3

Instinct and the Solitary Insect

THE COCOONS WE EXAMINED in the last chapter are silk-based structures that protect the developing young as they grow. An insect's preoccupation with the safety of the next generation finds expression in various other less elegant media: mud, for example. The risks to developing offspring go far beyond mere exposure to the elements or drying out. Insect eggs are highly nutritious, and are therefore prized by other insects, as well as by birds. Larval forms, especially caterpillars, are the particular target of many specialist species of wasps. And the food that wasps provide for their progeny may well wind up feeding the larvae of parasitic wasps and flies instead.

Without proper care, food stores may rot during the larva's week or so of development; fungi and bacteria are quick to spoil unrefrigerated liquids and solids alike. Or a parasite may lay its egg directly on the growing larva. Pupae, too, are at risk from a variety of predators, despite their shell-like cases.

Insects usually employ one of several strategies in providing for their young. The most common is to lay the eggs directly on the food—a plant, fruit, dung, or whatever—and leave the larvae that hatch to get on with it. This is typical of butterflies and many true flies. Others, as we have seen, provide shelter and camouflage for the offspring by bringing food to a nursery chamber and laying an egg there. The two variants of this more interventionist scheme are mass provisioning and progressive provisioning.

Mass provisioners such as the organ-pipe wasp are solitary. They build a chamber, fill it with enough food for the larva, lay an egg, and seal it off. Then they start another nursery. As we've seen, the major cognitive challenge here for some species may simply be keeping track of locations. For others, though, the necessity of choosing nest sites and the ability to adapt the nest to immediate contingencies, and to repair any later damage, suggest something more.

The strategy of progressive provisioning is more intellectually challenging. These insects maintain several chambers simultaneously, and the young may range from unhatched eggs through larvae ready to pupate. The amount of food required will certainly differ from one chamber to the next as the developing young may be either newly hatched or large and ravenous. The advantage of progressive provisioning is that spoilage of the food is not a worry: the next meal is delivered fresh at least once every day.

Progressive provisioners, if they put their nursery chambers in one place, may still be laying eggs or tending larvae when the first of their pupae hatch. A genetic curiosity in the Hymenoptera makes daughters more closely related to each other than to their own offspring, so the newly fledged wasps may be better off staying at home to help their mother rear more of her daughters (their sisters) than to begin new nests of their own. This mutually advantageous arrangement is the primitive beginning of sociality in insects such as ants, hornets, and honey bees. We will look at their coöperative building and social-coördination accomplishments in the next chapter; first we will examine the quite remarkable tactics invented by the single parents of the insect world, and the likely neural bases of their behavior.

RECYCLING BODY PARTS

The chitinous exoskeletons of arthropods allowed these animals to be the first out of the water and onto the land; they now dominate most of the earth's terrestrial niches. Chitin is indigestible except to some fungi, and it holds in water, a factor essential for life on land. It provides tough structures out of which the first strong jaws and proto teeth were made, the first legs able to support weight without the buoyancy provided by water, and the first strong joints. These skeletal components act as levers pulled by muscles; no longer did animals have to move and eat by squeezing internal compartments, protoplasm-filled sausages of incompressible but deformable cells. Arthropods can also grow a range of tools worthy of a Swiss Army knife: horns, antenna cleaners, pollen baskets, needles and stingers, wings, gyroscopes, saws, awls, clamps and pliers, hooks, and so on.

But chitin has its share of problems. For one thing, it does not stretch. To grow, the animal trapped inside its unyielding body armor must shed it and synthesize a larger outfit. Each change of skin also requires a period of helplessness while the new layer of chitin hardens. And chitin is metabolically expensive to make, far more costly than cellulose, the plant skeletal equivalent. Since chitin must be synthesized out of the animal's nutritional stores, selection will favor any ability to reuse some of this costly armor.

The pioneering naturalist William Beebe discovered two remarkable ways moths recycle their chitin in building projects. Both involve defenses against ants. The females of one species lay a circular sheet of eggs, and then set to work constructing a stockade to protect them out of their wing scales, which are long and narrow. Working at night, the moth weaves the scales into a picket fence about three layers thick, held together with silk. Hungry ants circle the palisade, eager but helpless to climb the slippery wall, and continue their search for food elsewhere.

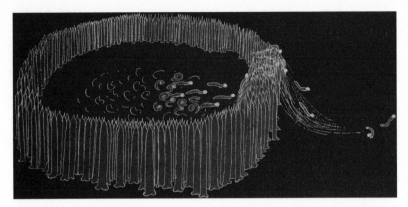

Palisade moth. The female moth lays her eggs and then removes the scales from her wings one by one; she uses them to erect a three-picket-deep circular fence on a leaf, attaching the scales to the leaf and to each other with silk. She then flies away. When the larvae hatch, they climb over the palisade and disperse to feed on leaves.

When the first eggs hatch, the tiny larvae crawl toward the light, leaving a trail of silk behind them. The larvae climb up the walls and out, each leaving a strand behind. By the time the last caterpillars crawl out, there is a substantial highway of silk to ease the journey. The palisade expands one important component of the species' niche: the range of possible nesting sites.

Clever as this construction looks, however, there is good reason to doubt that it is goal oriented. Beebe found that a slight discontinuity in the leaf could frustrate the builder. The moth, who could have started a fraction of an inch to the side and so avoided problems, instead works systematically around her circle of outer defenses; when she comes to a bump in the leaf, she stops erecting pickets. She searches along the projected path of the circle until she encounters a flat region of leaf, and then begins her fence again. She never tries the obvious tactic of continuing the fort up and over or inside the bump or hole, closer to the eggs, or of starting the project slightly to one side. Ants, of course, have no problem now; the eggs are easy prey. What is perhaps more notable is that other members of the genus do not build anything at all, which leaves us to wonder how this elaborate behavior could have evolved.

So the moth can have no picture of its goal, nor any understanding of the labor's purpose, despite the structure's seemingly intelligent design. Here is a simple example of programming with no alternatives to deal with common contingencies. The creature lacks the kind of third-tier map that would provide a sense of the relative position of things just out of reach in the world around it as well as the corresponding opportunity for selection to refine the behavior. Apparently the rote performance works so well so often that there is little pressure or reward for neural experimentation.

Beebe's other moth is in the business of guarding itself rather than its eggs. A number of species weave their toxic larval hairs into the cocoon. One of them even builds a spectacular geodesic dome, open except for the silk-wrapped hairs that form the lattice; the cocoon is slung inside like a hammock. But Beebe's caterpillar puts its stinging hairs to more direct defensive use.

When it is ready to pupate, the caterpillar finds a suitable branch and faces up a sloping twig. It then begins building the first of several spiny whorls. It plucks out a hair and connects it with silk to the twig. It anchors the next hair about 90 degrees from the first. After the first circle of four is built, the larva continues placing additional hairs at intermediate angles until there is no room for more.

The insect then turns around, moves down the twig, and builds a whorl to defend itself against attacks from the rear. This completed, the caterpillar turns again and throws up another array behind the first. By the time all the hairs are gone, there are generally four sets of radial spikes at each end. Finally, the creature can pupate safely; a week or two later, a clear-wing moth emerges.

Once again, we can be reasonably sure that nothing very brilliant is going on. Most obviously, this elaborate and painstaking defense is begun with little attention to the strategic situation. It is a fact of nature that twigs branch. If an animal builds at a Y-junction, it will

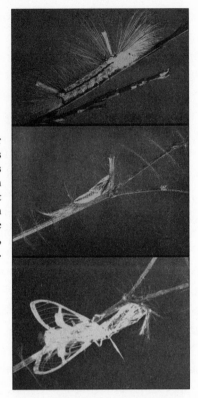

Whorl moth. The caterpillar of this clear-wing species removes the stinging bristles and affixes them with silk in a series of whorls in front and behind. The moth then pupates within this defensive structure, metamorphoses, and emerges as an adult.

need to defend against three avenues of attack rather than two. It would seem a simple matter to avoid such spots, but the caterpillars appear to be clueless on this essential point.

Worse, having chosen to build at a junction, the larva throws up whorls on only two of the incoming stems rather than on all three. Unless one set of toxic hairs happens to overlap the unprotected branch of the twig, predators will have little difficulty in reaching the helpless pupa. Here is a second example in which apparently simple backup programming for a common critical situation is just missing. And once again, no closely related species does anything remotely like this. The intricate performance that gives rise to the high-security fence seems to have evolved *de novo*, though it must depend in large part on a reordering of motor programs used in other contexts. One message seems to be that innately guided building behavior can change fairly dramatically under the influence of natural selection. Another is that these two insects have no picture of the world outside.

ROTE LEARNING IN HUNTING WASPS

Many moths lead intellectually unchallenging lives; with no special home to return to, nor any regular feeding place, the need for a map sense doesn't arise. Sphinx moths are a singular exception: some of these have home ranges, and they patrol a regular circuit. But the need for the spatial sense that is so crucial to many of the species we have looked at is not an evolutionary necessity for most moths.

Hunting wasps such as organ-pipe builders are in quite a different position. They have a home to get back to, and perhaps the need and ability to learn about good hunting sites. Tinbergen studied the way wasps locate their burrows, which are often inconspicuous. Some are carefully camouflaged by the wasp on her way out to stalk prey, perhaps with a stone fitted into the opening and dirt fanned over the stone. Given the poor vision of the hunters themselves, Tinbergen wondered how they manage to locate their well-hidden homes.

Their first problem is to navigate back to the general region of the nest; for this they seem to use the same methods that honey bees employ. This means that they take the sun as a compass or, when it is hidden by clouds or vegetation, the sun-centered patterns of polarized ultraviolet light. As we mentioned in the first chapter, this is no mean feat: The sun moves across the sky in a way unique to the date and latitude, and the wasp must learn this. But there are additional complications we did not touch upon.

Like sailors allowing for the current, flying wasps must also take wind into account. Headwinds and tailwinds will distort measurements of distance. If it is flying crosswind, the wasp must turn into the wind; thus the direction actually flown with respect to her celestial compass will be different from its orientation during flight. All in all, this is not a simple problem, even for a human. With her blurry vision, errors that take the wasp out of the range of familiar landmarks could be fatal. And, of course, learning the relative position of trees and other terrestrial guideposts for later use is challenging. In the next chapter, we will look in more detail at how this challenge is overcome.

Taking for granted that the wasp can get close, how does she find the nest she took such pains to construct and conceal? Tinbergen showed that the wasp triangulates its location on the basis of nearby landmarks, and that it learns the landmarks only on departure. Using a variant of the two-choice egg test he later tried on gulls, he worked out which features of landmarks are probably most important. Tinbergen laid alternating shapes, such as flat disks and smaller spheres, in a rosette around the nest while the owner was inside. When leaving, the wasp would hover near the nest, apparently studying what she saw. On her return, she found two circles of landmarks, one of disks and the other of spheres. The wasp always flew to the center of the three-dimensional cues, but remembered the disks as well: given a circle of disks and another of unfamiliar sticks, she chose the discs she had previously slighted. These insects appear to have a local third-tier map that allows them to plot nearby landmarks. They then use this information by flying about near the nest and comparing the remembered picture of the landmark array with what they are seeing. When they have a match, they are above the entrance to the nest.

So hunting wasps have a good memory for landmarks and their location both near the nest and far away, and they know the relative locations of these various markers as well. Among wasps that are progressive provisioners, the powers of memory are even greater. One of Tinbergen's students studied a wasp that preys on caterpillars. A female may maintain as many as a dozen nests at once, each carefully hidden. Already, with several scattered homes, the memory load is much greater than is needed for most species, consisting as it must of multiple third-tier maps. But there is more: the wasp needs to remember where the burrows are relative to one another. This probably involves a larger scale fourth-tier system.

The cognitive load on these hunters can be greater still. The wasp's several burrows may contain larvae of different ages and sizes, and therefore appetites. The number of caterpillars needed to feed each depends on the age of the growing larva, and whether it is a male, which will not grow as large, or a female. At the beginning of each work day, the female visits the burrows one after another, apparently noting their needs. Larvae ready to pupate are sealed in; the rest must be fed according to their appetites and level of development, and a new burrow or two begun. The wasp sets out with a mental shopping list, and delivers to each nest the number of caterpillars indicated by her initial survey. If researchers substitute young larvae for old and vice versa, the wasp takes no notice: if the burrow held a small grub in the morning, it gets only one caterpillar, no matter how large and ravenous the occupant is when the wasp returns.

In some ways, the very complexity of the task—keeping track of the locations of a dozen burrows and the changing needs of their occupants—may demand that it be performed by rote. Most humans find it hard to remember so many things without a written list. At least at the level of the wasp, the ability to memorize many things does not seem to indicate great mental powers; the difficult has been made easy with specialized wiring dedicated to this daily task. We will look at this question again when we examine honey bees, one of the best learners in the animal world.

We can entertain doubts that the extent of memorization ability implies any degree of comprehension, but does the spatial ability of hunting wasps translate into some more goal-oriented facility in building? The wasps can certainly learn in some contexts, but can they bring

Cricket ritual. This species of hunting wasp specializes on crickets, which are paralyzed with a precise set of injections into ganglia along the ventral midline. She trims the cricket's antennae to a convenient length, and sets off home. When the wasp reaches the burrow she has prepared, she leaves the cricket about an inch away and inspects the nest to check for parasites. She reëmerges, takes hold of the stubs of the antennae, and pulls the prey back into the burrow. Moving the cricket while the wasp is underground forces her to realign the paralyzed prey and reinspect the nest.

learning to bear on their architectural challenges? The evidence for some minimal understanding of nest construction by solitary wasps is uninspiring. Many of the first astute observations were made by J. H. Fabre, a teacher who studied insects during term breaks. It was Fabre who showed that mass provisioning wasps paralyze rather than kill their prey, and that they specialize on particular species.

Fabre found the burrow-provisioning ritual of these hunters fascinating. A species that takes crickets, for example, carries the prey under its body, head first, to within an inch of the burrow. The wasp releases the paralyzed cricket, its head facing the opening, and goes into the tunnel, possibly to check for parasitic eggs or larvae. She returns to the surface, grasps a cricket antenna in each of her front feet, and drags it down into the burrow to serve as a meal for the hatchling.

If Fabre rotated the cricket even 45 degrees while the hunter was below the surface, the wasp would be visibly upset. She would reorient the cricket with great ado, and then reinspect her underground works. While she was below, Fabre might move the cricket an extra inch from the opening. The returning wasp would seem even more annoyed, return the prey to the "right" spot, and then disappear for another inspection. Fabre generally gave up after two dozen of these cycles; the wasps almost inevitably outlasted him.

Another example bears more directly on the wasp's architectural understanding. Fabre replaced the temporary closing of the tunnel with a wholly impenetrable plug. The returning wasp, arriving with food for a developing larva underground, would become consumed with this novel problem. She would dig, search, dig some more, fly around to be sure she was in the right place, dig some more, and so on. Fabre then excavated a trench that gave the wasp direct access to her hungry offspring. But the insect walked back and forth, actually treading on the larva in her desperate search for the opening. Even when the goal was underfoot, she could not shortcircuit the sequence. She seemed to have a very local picture of the nest location, but not of the nest's structure.

Fabre believed that all insect behavior is innate, and he dismissed the abilities of wasps with a typically pithy comment: "The insect which astounds us, which terrifies us with its extraordinary intelligence surprises us, at the next moment, when confronted with some simple fact that happens to lie outside its ordinary practice." But as we have said, intelligent behavior can involve orchestrating innate components; as a result, simply categorizing behaviors as innate or learned doesn't really tell us very much, and distracts us from the goal of understanding animal minds. We need to see how these elements work together. Perhaps the best-understood species in this regard is the so-called funnel wasp.

FUNNEL WASPS

A number of solitary wasp species build above-ground structures. Even though different species construct an extraordinary variety of nests, most are built with similar strategies for guiding motor programs and judging progress. In many ways, the most interesting of

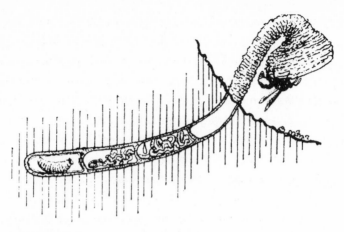

Funnel wasp. This species of Australian wasp excavates a tunnel, lines it with mud, then builds a funnel to protect against parasites. The funnel consists of a stem roughly perpendicular to the ground, a curving neck that brings the opening to about 45 degrees from the horizontal, a bell that widens the opening, and then the funnel proper. Once she has completed it, the wasp gathers prey (caterpillars), paralyzes them, and when there are enough, lays an egg and seals off a section of the tunnel. The egg hatches into a larva, which eats the caterpillars and then pupates. Meanwhile, the female wasp completes more cells, and finally tears down the funnel and covers the opening with soil. In this sketch, the oldest (deepest) cell contains a pupa, the next a larva feeding on prey, and the shallowest an unhatched egg among paralyzed caterpillars. The female is beginning to tear down the funnel.

these is a hunting wasp from Australia that excavates a burrow, lines it with mud, fabricates a protective funnel (also out of mud), and then hunts for caterpillars.

Like other solitary insects, these *Paralastor* wasps build in a strictly linear order; there is no multitasking, no working simultaneously on separate parts of the structure. They construct the burrow, then the straight tube of the funnel, then the curving neck, followed by the widening flange, after which the graceful bell is made. The structure is rough on the outside, but the interior surface is almost slippery. Its purpose is to keep parasitic wasps from getting into the burrow and laying eggs. The wasp gains access by landing on the bottom edge of the bell and stretching her front legs until she gets a purchase inside the neck; she then pulls herself in. Since parasites tend to be smaller than their hosts, a bell that makes it difficult for the hunting wasp to get in is likely to be too much of a challenge for a parasite.

If the animal were a computer, its construction program would consist of instructions (motor programs) for each step; the animal would run through the steps again and again, with signals (sign stimuli) the wasps would look for after each iteration of the motor program. When some innately recognized criterion was reached, the insect would pass to the next set of motor programs—from extending the stem to fabricating the neck, for instance. But even if a builder works from such an automatic strategy, it might still have some more general understanding and flexibility, some third-tier picture of the structure being built. Selection would favor a default plan that would carry the behavior through its normal course in any event. An ability to deal with contingencies, especially unlikely setbacks, could suggest something more.

Workers have dissected the wasp's funnel-building sequence with an imaginative combination of observation and experimentation. The first part occurs underground, so the details cannot be directly observed. The female excavates a tunnel, generally into sloping ground or a bank. Once she has removed the dirt from a narrow cylinder two inches deep, she coats it with a smooth layer of mud. The hardened tunnel lining keeps the interior dry and safe from cave-ins.

The stem comes next, built perpendicular to the ground. From a vertical bank it extends out horizontally; from a horizontal surface, the stem projects straight up. This strategy serves to place the funnel as far from the ground as possible. But though the angle of the stem varies according to the angle of the ground, the orientation of the funnel components constructed later is the same in all nests.

Stem angle versus funnel orientation. The angle of the stem depends on the ground; it is approximately perpendicular to the surrounding soil. The orientation of the bell and funnel, on the other hand, is constant. The length of the neck depends on how much curvature is needed to continue the cylindrical entrance to the angle needed for starting the bell.

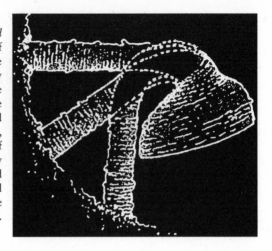

The building cycle involves flying off to a damp place to collect mud, bringing it back, and working it round and round to extend the cylinder, whether it be tunnel, stem, flange, or bell. Three simple experiments show how the mind of the builder operates during this stage of the work. First, we can reorient the stem—that is, snap it off at the base, and glue it back at a different angle, leaving of necessity a wedge-shaped gap at the bottom. The wasp, it turns out, is committed to building along the axis of the existing tube; once the stem is begun, she pays no further attention to the ground angle, and makes no attempt to fill in the gap. Thus, if the tube had been vertical and is now attached at an angle of 45 degrees, the insect will continue extending it along this new diagonal line, ignoring the missing wedge at the base.

Construction of the stem continues until it has reached a predetermined length; then bell formation begins. This building program is sufficiently automatic that when experimenters partially bury the stem during construction, the wasp keeps extending it. But once she has extended it far enough to begin the curving neck, she is committed. Adding more soil has no effect, and she will even build the funnel right down onto the ground, defeating the purpose of the whole exercise.

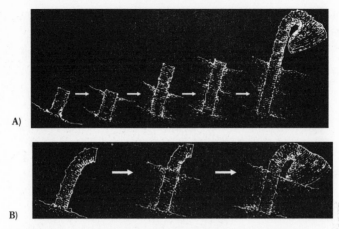

Stem-height test. (A) The stem is built until the exposed cylinder reaches a certain height. If the ongoing construction is buried, the building continues until the length criterion is reached, which triggers the next step in the process. (B) Once the neck is started, burying the stem has no effect. The wasp will build the funnel down into the ground, defeating the purpose of the structure.

Another test reveals how the wasp knows when the stem is tall enough. As we saw with caterpillars, and shall see again when we look at birds, a common tactic for an animal is to use itself as a mold, an interior form around which the nest is built, just as shoes are constructed around wooden lasts. Thus the nest that is constructed either is just the size of the sitting bird or has elements just as far away as the bird standing on a perch can reach. This is what the wasp does: when, standing on the ground, she can just reach the top of the stem, this first element of the funnel is exactly the right size. Adult wasps of the same species can vary fairly dramatically in size, depending among other things on the amount of food that was available when the wasp was a larva. If we compare the structures built by four *Paralastor* females ranging in size over almost a factor of two, we find that each builds stems of a consistent size, but the smallest stops when the base is an inch or less high, and the largest turns out stems one and three-quarters of an inch long.

Stem height table. The stem height criterion depends on the size of the wasp. Larger wasps reliably build longer stems (and deeper funnels), using themselves as the measuring stick.

Wasp:	A	B	C	D
Experiment: 1	25	30	32	44
2	26	30	36	42
3	23	30	32	43
Average:	24.7	30.0	33.3	43.0

Once the stem is finished the wasp begins the neck, which is an arc with an effective diameter of about three-quarters of an inch. If the stem is horizontal, only a small bit of neck is needed: once the "tall-as-me" criterion for stem height is reached, neck construction continues from the vertical opening of the stem only until the angle of the bit of stem being built reaches 45 degrees from the horizontal, facing down. If the stem is vertical, on the other hand, the neck curves through 135 degrees. And as it was with the stem, so it is with the neck; the wasp works until she senses the stop signal, and only then moves on to the next step, never to return to the previous job. Thus it is that the neck begun on a stem that was horizontal can be extended by presenting the returning wasp with a neck on a vertical stem.

You can continue to rotate the stem so the builder never achieves her 45-degree goal, and the result is an endless helix as the wasp strives fruitlessly to reach the target angle. But once she senses that the criterion has been met, she moves on to build the flange; changing the stem angle now has no effect. It can be rotated so that the funnel will be built horizontally, vertically down, or even straight up. Having committed to the flange, she will not turn back.

Once the flange is finished, the wasp extends the bell until she can just reach the neck while gripping the bottom of the funnel. Should she add even another circle of mud, she would be unable to get into her own nest. The actual size, of necessity, depends on the length of the builder. Like the stem, neck, and flange, the outside is rough, but the interior is polished smooth; no parasite is able to crawl into the neck. Now is the time to hunt for caterpillars, stock a cell, lay an egg, add a compartment wall, and then repeat the process until three larvae have been installed. Her final act is to raze the elaborate funnel and camouflage the nest.

The construction behavior of *Paralastor* females is simply but reliably controlled. In the absence of human intervention, a set of motor programs cycles through until a criterion is reached, and then the wasp moves on to the next set. Whether it is excavation, lining the tunnel, fabricating the stem, constructing the neck, building the flange, finishing the bell, or collecting caterpillars, each process continues as long as it is needed, and no longer.

It is clear that the funnel wasp has no third-tier picture of the intended finished product. Any of a variety of unnatural modifications can lead to absurd behavior, as we saw when the stem was buried or reoriented. A seemingly trivial hole drilled in the neck baffles the builder, and usually induces her to start an entire second funnel where a small patch job would have sufficed.

As far as we know, the same pattern holds true for all solitary wasps. Although the hunting wasp knows enough about the locale to navigate, enough about her nest entrance to find it, and (for the progressive provisioners) even the details of up to a dozen separate burrows, the construction behavior proceeds without any larger understanding. The builder cycles through a behavior until she detects a cue signaling completion, then moves on to the next step. Backtracking and repair are impossible.

It could be that this linear and hierarchical system of the funnel-building wasp is so reliable under ordinary circumstances that a greater flexibility would sacrifice speed and certainty. What does seem clear is that the behavior can be quite compartmentalized: in one context a wasp may seem smart and flexible, but in another it appears to be a rigid robot. The ability to learn and plan routes is irrelevant; there appears to be no transfer of these fourth-tier cognitive skills into the domain of animal architecture. When it comes to building, solitary insects provide textbook examples of innate control.

A more important point, however, is that here again we see the pattern of dramatically different building behavior evolving in one species while the other members of its genus continue with conventional unadorned tunneling. Architectural breakthroughs are possible, and need require no insight; but most often they are evolutionary dead ends, too specialized to admit of useful adumbrations. In the next two chapters, we will look at cases in which innovations in design and construction have so dramatically widened the niche of an ancestral species that it has radiated into dozens, or even thousands, of new species, exploiting variations of the original novelty. This is a familiar pattern in other aspects of evolution: the ability to digest cellulose allowed the explosive diversification of fungi; the invention of flight by early birds opened up thousands of aerial niches; fur, the womb, and mammary glands were a winning combination for the ancestors of placental mammals.

Rampant architectural-based speciation is especially obvious in social insects. And with social life came an entirely new and demanding building strategy, complete with its own cognitive consequences. Any structure may be added to by another member of the colony while the original architect is off gathering new material. For individuals of such species, flexibility is essential; builders must always be able to switch to any stage of any project. At the same time, each insect must adhere to some more general plan specifying colony-wide construction strategy—angles and distances, gradations and curves. As we shall see, the evidence for goal-oriented behavior in these creatures, often near relatives of the less intellectually inclined hunting wasps, is strong.

CHAPTER 4

Social Intelligence of Wasps and Ants

SOCIAL INSECTS REPRESENT the apex of invertebrate evolution. Some species live in communities of millions, coördinating their building and foraging, their reproduction, and their offspring care. Yet sociality is found in only a few species of insects. Among vertebrates, sociality is rare too: wildebeest and lions are the exception rather than the rule. Nearly all fish, amphibians, reptiles, birds, and mammals are solitary, except when courting and mating. Birds and mammals usually rear their young, but year-round family groups are almost unknown, though they are intensely studied where they do exist. The same is true for insects.

We know, or think we know, that social groups are good things. Wolves are better predators when they hunt in packs; pigeons escape from falcons far more often when feeding in flocks; and group building projects—the dams and lodges of beavers, for instance—can provide a high level of protection and comfort. Why, then, are social species so very rare? In fact, living socially presents inevitable problems that transcend niche and habitat. Only when these costs are amortized by corresponding benefits is group living a plus.

The most obvious cost is competition. All the members of a species share the same niche; when they live together, they are trying to eat the same food and occupy the same nesting sites. In general, there is far less competition away from a group, and selection should favor any individual who (all things being equal) sets off on its own, leaving the members of its group behind to compete among themselves for limited resources.

Another difficulty is that concentrations of individuals facilitate disease and parasite transmission. On the whole, social animals carry more parasites and species-specific diseases than their solitary brethren. Parasites and diseases sap the strength and stunt the growth of animals, and among highly social creatures, epidemics can devastate whole populations. Whether it is distemper wiping out entire colonies of seals, the Black Death killing a third of Europe's humans, tracheal mites dooming half the colonies of honey bees in the United States, or Old World diseases condemning more than half of Native Americans to early graves, the penalty of social life is potentially huge.

But if the niche is right, the payoffs can be even greater. We've already mentioned two: coöperative hunting, and defensive groups. Social hunting is likely to evolve where prey is too large to be taken by individuals operating alone. To capture wildebeest, some members of a lion pride stalk the herd and flush them toward others lying in ambush. In other species, individuals forage or hunt simultaneously and share the food. Vampire bats that have had a bad night, for instance, are fed by more successful members of the community, but they are expected to return the favor in future. Coöperation can even involve sharing information about the location of food. Some colonial birds, such as bank swallows, use the departure direction of a successful forager to locate concentrations of prey. This information transfer can be unintentional or, less often, make use of special assembly calls or behavior.

Coöperation in group defense (such as we see in circles of musk oxen or elephants), is quite rare among vertebrates, but is prevalent among the social insects. The strategy of employing many eyes to watch for danger, on the other hand, is widespread in birds and mammals. A herd of gazelles is far more likely to spot a lurking lion or concealed cheetah than is a lone individual, and at a greater distance. In fact, a group enters into a kind of time-sharing arrangement in which individual antelope alternate biting off a mouthful of grass with a period of erect and watchful chewing. A larger group can afford more bites per individual per minute, there being more eyes to scan for danger. If you are a small antelope living in a forest where visibility is limited, however, remaining hidden is probably a better bet than assembling into noisy herds.

Among the millions of species of insects only a few thousand are social, and those rarities are confined to the termites and Hymenoptera. All termites are social; as we shall see, this is because their diet (cellulose) requires that each generation feed a special kind of bacteria or fungi to the next—one of the few instances of a beneficial social disease. The hymenopterans are more numerous, comprising sawflies, wasps, ants, and bees. Sawflies (one word to entomologists because they are not true flies) are never social, but all ants are. Bees and wasps come in both varieties, though solitary is far more common. Ants and bees evolved from wasps, a hugely diverse group. Except for their winged reproductives, ants are basically wingless and stingless wasps. Bees are vegetarian wasps; they get their protein from pollen rather than by killing or paralyzing other animals.

Where evolutionary chance and ecological circumstance have permitted, sociality has developed. As we mentioned in an earlier chapter, a genetic quirk of the Hymenoptera makes sisters more related to one another than to their own offspring. Biologists like to talk about "true sociality" (eusociality), by which they mean there is a caste of nonreproductive workers who sacrifice their own chances to have offspring in order to rear siblings. This so-called altruistic selfishness is also seen in one group of vertebrates, the mole rats of

Africa. Among insects it has evolved independently more than a dozen times, once among roaches (giving rise to termites), and on at least eleven occasions in the Hymenoptera. (There are some other species of insects that are technically eusocial, but which have little or no organized social structure.)

There is a trend in psychology to differentiate general intelligence into specific forms, one of which is the hot idea of social intelligence. The theory is that social living makes possible—indeed, requires—a new plane of mental activity. Many societies, for instance, have an ingrained hierarchy; members must be able to recognize each other as individuals, and each animal must keep track of its rank relative to others for the community to succeed. This hierarchy is a kind of linear map, a cognitive challenge that nonsocial animals or ones that live in very small groups simply do not face. Social grouping can present other situations that require more flexible mental processing. If a group is coöperating to build a structure, for instance, each animal must be able to work on whatever step is under way, and at any stage of completion. When the society employs a division of labor, each individual must decide what most needs doing at the moment, all the while taking into account the general "investment" pattern of the rest of the group.

The view that social life requires more and different intellectual ability than solitary living seems entirely reasonable. However, mole rats are rarely thought of as more intellectually capable than other mammals—humans, for instance. Nevertheless, there does seem to be a pretty broad correlation among insects between social and smart. As we shall see, there is a complex interplay between niche, architecture, sociality, speciation, and cognition; quite possibly this is a positive-feedback loop that has amplified the benefits, driven innovation, raised the intellectual stakes, and led to enormous diversity. We will look at the architectural achievements of all degrees of sociality, from casual cohabitation by reproductively capable females to the irreversible celibacy of worker ants, keeping an eye out for the likely cause-and-effect evolution that makes hornets, honey bees, and termites seem more cognitively complex than digger wasps, carpenter bees, and roaches, their solitary counterparts.

THE ROAD TO SOCIALITY

The ancestral hymenopteran was a wasp-like insect with the characteristic four pairs of wings, long antennae, and rapid flight that persist today. Primitive forms laid their eggs in or on fruit, in galls on stems, or on the larvae of other insects. There was no offspring care, nor any building beyond a cocoon. The first step toward sociality was architectural: some species began building protective structures for their larvae such as the burrows described in the last chapter. All food was provided before the egg was laid. Selection will favor this move only if the time and effort expended in construction are more than repaid by a higher survival rate among the young. The change, when it occurs, alters the species' niche. Indeed, every step toward sociality impacts the niche itself, giving rise to new opportunities and novel challenges. For those that also evolved a third-tier mapping ability, new intellectual tools became available.

The next logical move in protecting developing offspring was for the parent to guard them from harm during development. Again, this requires adults to sacrifice energy and feeding time; only where predation and parasitism are major threats, and a parent on site can deter them, will selection reward guarding. In fact, most of the parasites are freeloading wasp species that capitalize on the hunting and provisioning of other wasps. In common justice, some of these parasites are themselves targeted by even smaller species of parasitic wasps (which include some of the most minute of all insects).

Guarding isn't always a full-time job. For some kinds of burrows, and in certain habitats, the nest need be checked only a few times a day at most to defeat parasites and predators. This is particularly true of burrows having well-camouflaged entrances, whether they be in the soil, in trees, or under leaves. For these, selection might favor some kind of time-sharing; the female may begin new nurseries while larvae and pupae mature in the older ones. This works if the increased risk to the young from part-time security is amortized by the enhanced reproductive potential arising from starting more nests. It requires an ability to store and use multiple maps, another likely instance of the value of neural duplication.

Once some species had adopted this tactic, the next potential improvement was progressive provisioning. By feeding the larvae only what they need each day, the wasp need not paralyze the prey as a preservation measure. Since accurate placement of venom injected into the nerve ganglia is essential for paralysis, most mass provisioners specialize on one particular species. They know the location of neural control centers and how to reach them through the weak point in the body armor of one, and only one, kind of victim. A progressive provisioner, on the other hand, can use anything it can capture and kill because the prey will be eaten within hours or even minutes. For these wasps the range of food species is huge, and thus the niche is correspondingly wider.

If the nurseries are clustered—as branches off a single tunnel, for instance—then when the young of progressive provisioners emerge, their guarding mother may still be there. For a few species, the daughters that hatch out may coöperate with the mother to enlarge the nest and rear another generation. Although the unusual genetics of the Hymenoptera make this potentially advantageous, there must be new control systems to ensure that all tasks are done, that not everyone is building or collecting food or tending brood at the same time. Unless the many essential jobs are parceled out in relation to the colony's needs, chaos will reign. A rote sequence of duties must be replaced with cueing from local stimuli, but with additional controls to maintain a mix of ongoing work. Presently, we will look at how this new level of social control operates.

Although wasp colonies can begin with a single queen engaged in progressive provisioning, some quite common ones start with multiple fertile females. This is often an example of propinquity: reproductives of a species are all looking for the same kind of place to rear their young, and they wind up building side by side. This situation produces friction and, when egg laying begins, overt fighting. The largest or first to arrive tends to dominate the others, eating the eggs the rest lay or the larvae that develop from them. Some less dominant individuals may leave and start over on their own, but others (generally sisters of the alpha female) remain and help. Kinship recognition depends on odors that have genetic and individual components. The consumption of eggs has evolved into extraordinary feeding ceremonies in many highly social species; a worker ant or bee or wasp may produce a nutritious but infertile egg and then, while other workers crowd around attentively, present it to the queen to eat.

In the northern temperate zone—the United States, southern Canada, and Europe in particular—the most common semisocial wasps are the Polistes. Their annual cycle begins in April or May, depending on climate. Females mate in the fall and overwinter under bark, in leaf litter, or in other protected, dry, and insulated spots. Choosing where to spend the cold months is another cost/benefit decision: the best answer depends on the unknowable severity of the winter ahead and the range of unoccupied sites available. A location that is exposed enough to provide early warmth, and thus to allow a female to start work as soon as possible in the spring, may be one that spells certain death in an unusually cold winter. Females might search and choose at random, or they might weigh the several factors under their control and look for an optimum combination. We do not know which tactic they employ, but we will see that in other social species this decision involves considerable research and evaluation; a similar sophistication may be involved here. It is surprising, though, that the reproductives of these species never, so far as we know, build winter shelters.

For the Polistes wasps, as for similar species, the need to work out the best combination of factors arises again when the queens that survive select a site for building. The spot must be protected from rain and sun; ideally, it will be concealed from predators and insulated from rapid temperature changes. The queen constructs a thin, extremely strong stem—a pedicel—out from a support surface, and then generally coats this strand with chemicals that deter ants. The pedicel is manufactured from plant fibers the wasp has stripped from stems, chewed with water, and then drawn out in such a way as to bring the stringy fibers into alignment.

Upon this seemingly precarious foundation she begins a piece of comb, but only about seven cells at first. The comb is fabricated from paper, which she makes by scraping the surface material from twigs and then chewing the mass with water. Armed with a ball of damp cellulose, she works on each cell with just the kind of circular motion so evident in the funnel wasps; but here the builder uses her mandibles to form the thinnest possible layer. When dry, the result is a paper of remarkable strength. The female divides her time between building cells and gathering food to feed the seven larvae that have emerged from eggs laid in the cells.

In a successful colony, there will be several overwintered females building and provisioning together; within a few weeks one wasp will be dominant, and will specialize in egg laying. The number of cells grows to perhaps 250 by midsummer, and with it the number of sterile workers. As summer progresses, the efforts of the colony are redirected more and more toward producing a population of reproductives (male drones and large potentially fertile females). As these high-caste offspring do no work beyond incidental guarding and some personal foraging for nectar, the colony works more and more at a loss, catching prey, pre-chewing it, and feeding it directly to this elite caste. By the end of the summer, the members of the ever-growing reproductive group—the only chance the workers have of perpetuating their genes—are so numerous that developing

Paper wasp comb. This open nest species builds about a
hundred hexagonal paper cells supported on a pedicel. By
the end of summer, most of the cells are being used to
rear reproductives rather than workers.

larvae and pupae may be torn from the comb and fed to them. Mating occurs in August, and by the end of September the fertile females
have abandoned the nest.

We will look more into the building behavior presently; for now,
we should consider how much coördination and social knowledge is
needed to manage this group enterprise. For one thing, a linear hierarchy exists among the ur-females that coöperate in nest construction and provisioning: the dominant (who beats up the others
and eats their eggs) is the alpha female. But careful observation reveals that there is a beta female as well, who dominates those below
her. The next in line is the gamma, and so on. The reality of this
progression of place, this order of entitlement to whatever may be at
issue, is especially clear when the alpha dies or is removed. The beta
takes over the egg laying, and everyone else moves up a notch.

Although we are used to sorting such chains of command in communities of baboons, herds of horses, and flocks of chickens, it seems
surprising among insects. A strict dominance hierarchy requires

Social 0:	Social isolation: conspecifics are either ignored or attacked.
Social 1:	Social hierarchy: animals have a linear representation of part or all of the social order, especially individuals ranked near the individual in question.
Social 2:	Decision-network mapping: multidimensional representation of parameters is important in making social choices.
Social 3:	Attribution and intention: animal has an ability to understand the cognitive processing in the brain of a conspecific, and can alter its behavior to exploit that knowledge.

individual recognition to keep each member of the society from fighting when it comes in contact with another. Hierarchies establish likely winners and losers; since each animal remembers its place, it is less likely to lose time or risk possible injury by contesting every bit of food or place to stand. However, there must have been major changes in programming for this behavioral switch to be successful; remembering one's place in a social matrix (based, in wasps, on facial memory) is a substantial cognitive task. It is this kind of social map (Social 1) in which individuals have some awareness of a social hierarchy, combined with the ability to shift jobs as needed and to pick up tasks at any arbitrary point between beginning and end, that must make us wonder just how much social insects comprehend about the larger fabric of their communal enterprise.

EUSOCIAL WASPS

The vespid wasps present a key example of the interaction of architecture and sociality. The group comprises three primitive subfamilies of solitary wasps that burrow or build with mud; they represent the modern descendents of the original vespids. From these there arose the Polistes, the subfamily we have been discussing, which began with the discovery of how to make paper and fashion it into nests. This breakthrough made sociality a plus, broadened the available range of niches, and led to the evolution of many new species. One of these new types invented the enclosed paper nest with a small, easily defended entrance; this innovation further widened the range of niches, generating substantial speciation, and led in turn to a much higher degree of sociality. This advance selected for even larger colonies. Two new groups spun off: the (mostly) temperate-zone vespids and the polybiine wasps (most common in the tropics).

Sociality has evolved to such extreme levels in ants as well as in some wasps and bees that it is no longer optional. These eusocial insects have no fighting for dominance, there being a reproductive female caste (generally a single queen), her numerous sterile daughters, and (as needed) reproductive males. Everyone is related; even if the queen mates more than once, the workers are all sisters and cousins and aunts. In many species the workers' ovaries are kept inactive by chemicals produced by the queen, known as queen substance. In one vespid wasp, the queen substance is:

Any chemist will notice at a glance that this molecule is too large to evaporate, and it has an oily character. The odor, then, is not in the air of the colony, but rather on the bodies and in the communal food; it is still detected by the antennae, however. We will look at this kind of chemically based social control a bit more closely in honey bees, where it is understood in some detail.

When a colony or, indeed, any society, grows past a certain point, individual recognition of all members of the group becomes impossible. In eusocial insects, individuals can distinguish the queen from all others, and various groups as classes: males, female reproductives, workers, and the larvae and pupae of each of these three groups. This is analogous to our initial categorization of other humans into babies, children, and adults, each subdivided into male or female. When there is no dominance order, there seems no reason for selection to favor more elaborate recognition. It is ironic, then, that a higher level of sociality may reduce the cognitive load in this regard. Only the colony odor need be learned; the other elements of olfactory recognition can be hardwired. But this is not quite true of some species: workers can distinguish finer gradations of kinship—full sisters from half sisters, for instance, still based on odor.

Though the caste system reduces counterproductive competition to a minimum, it also means that replacing a queen is a serious problem. In some species, such as ants and termites, it is impossible: when the queen dies or runs out of sperm, the colony starts to die. In others, given time, a new queen can be reared during reproductive season. Honey bees have the best chance; workers in their perennial hives can rear a new queen, and the mating season lasts all spring and summer.

Another advantage of being eusocial and building an enclosed home is climate control, which again has a huge impact on niche. In the temperate zone, social wasp colonies do not survive the winter. But during the milder part of the year, when they are active, the great number of workers can construct an insulated and waterproof nest. By opening up many more habitats and potential nesting sites, this architectural strategy increases the potential range ever farther from the tropics. It also greatly improves larval and pupal survival, since the young are not exposed to the sorts of extremes that cause developmental problems. The offspring also mature faster at the elevated temperatures maintained by group-generated structure and behavior, meaning that new workers are available in perhaps three weeks instead of four. This permits a conservative estimate of a nine-fold advantage in growth potential over the course of the season. And larger colonies can build larger nests, with better insulation, making possible still bigger societies that can survive even farther north, but which require more elaborate and flexible mechanisms of social communication.

Temperature regulation in the nest takes two forms: keeping things warm, and cooling them off. Both operate best when there are enough workers that some can focus on environmental engineering while others take care of foraging or rearing the brood. Cooling is evaporative: water is brought to the nest by a self-selected cohort of workers, spread on the comb, and fanned. Insulation not only minimizes temperature extremes as well as the drying and disruptive effects of wind, but it makes possible active heating as a subset of adults cluster on the comb and pulsate their wing muscles. The design of hymenopteran wings allows them to uncouple this kind of isometric exercise from wing movement, and the result is almost pure heat. The nests of eusocial wasps are kept within 3 degrees of 86°F; humidity variation is also minimized. Honey bees do even better than this, controlling the temperature of the brood to within 0.5 degrees, which is about as well as we do with our internal temperature.

The best insulation other than a vacuum is trapped air. Feathers and fur keep animals warm only to the extent that they prevent air from moving between the outside atmosphere and the surface of the skin. So the problem for temperate-zone paper wasps such as hornets is to create a structure that captures and holds air. Hornets do this by constructing a series of concentric paper spheres around the comb. These are easy to see in a young nest, but become less apparent as the colony grows.

A problem is instantly evident: how can a new colony possibly build a sphere large enough for the comb it will need when the population rises into the hundreds? As with human homes, the answer is renovation. The earlier inner layers are constantly being torn down by workers specializing in building, and then replaced by new ones on the outside. As the structure grows, however, an unreinforced sphere of sufficient size is impossible; long arcing chambers

Paper wasp nest under construction. This bald-faced hornet nest has several tiers of cells. As new, wider tiers are added, the wasps must remove one or more of the inner layers of insulating paper while constructing new sheets on the outside. Two wasps in this relatively young nest are working on new strips of paper, extending the layer toward the entrance.

are built instead. These are particularly evident at the top, where the first two layers of comb are eventually incorporated into a highly insulated, multichambered attic.

The spheres are quite attractive up close, where the different colors of paper are evident; each stripe represents the pulp load of one wasp worked into a thin layer. The pattern also reminds us how coöperative the building process is; an individual wasp visits the same pulp source again and again, but rarely does she return to find that she can take up work where she left off. Thus one wasp's stripe of yellow siding shifts abruptly to the brown carried by another, which gives way in its turn to gray. There is no way the rote funnel-wasp strategy could work here. The builder needs to know where she is in the overall structure under construction, and what needs to be built there. Sociality requires that the building behavior of individual wasps become far more flexible so that the wasp can switch from place to place, from task to task, from stage to stage. This is one aspect of the kind of task mapping that requires processing at the Social 2 level.

The utility of multiwall insulation seems to transcend locale. Although hornets generally build their paper spheres suspended from

Paper wasp nest detail. Each worker involved in exterior building adds a mouthful of pulp, which she works into paper. Every visit creates a stripe the color of the bark the wasp has harvested.

trees or in bushes, the European yellowjacket (now the most common form in the eastern United States) constructs exactly the same sort of nest underground, typically in an excavation abandoned the previous year by another species—most often mammals such as mice. Hornets may also elect to build underground, below eaves, or in a shed or attic. As a mated female searches for a suitable place to start a colony she is guided by a multitude of factors, including the cavity's volume, its camouflage, accessibility, and exposure to wind and rain, each of which must be weighed against the others. The behavior during searching (especially in honey bees, where it has been studied extensively) looks like third-tier mapping.

The internal structure of most wasp nests is distinctly counterintuitive: the comb is built in horizontal tiers, the openings facing down. Why don't the larvae fall out? One of the things that makes silk so strong is the countless number of weak electrostatic bonds between the atoms in adjacent chains. Some substances have a few of these polar atoms sticking out, ready to bond, some have many,

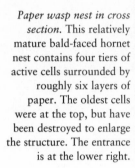

Paper wasp nest in cross section. This relatively mature bald-faced hornet nest contains four tiers of active cells surrounded by roughly six layers of paper. The oldest cells were at the top, but have been destroyed to enlarge the structure. The entrance is at the lower right.

and some (waxy and oily molecules) have none. Probably the most polar molecule in nature is water, which is 70 percent of the cellular cytoplasm. One of the most electrostatic of the organic compounds is cellulose. Wasp paper, which some historians think was the inspiration for the first human paper, is almost entirely cellulose.

As we all know, the affinity of water for uncoated paper is enormous. Water will climb readily up a paper towel against the pull of gravity. This affinity between two electrostatic substances is also the basis of capillary action—the capacity of water to ascend narrow glass (but not plastic) tubes. Larvae are damp, as is the food placed in their cells; they simply stick to the paper walls. When they are ready to stop moving and pupate, however, they have to keep from sticking; they do this by attaching themselves to the cell with silk, and then spinning a silk cocoon. Workers cap the cells with seals that the emerging adults will chew through from ten to fourteen days later.

The nest grows as each tier is widened, and new tiers are added. Access from one level to another is at the periphery, just inside the shell. Each tier is held by a strong pedicel in the center, and less sturdy supplemental supports between the layers somewhat farther from the center axis. Most of the building, therefore, is upside down and in the dark. Although the cells are hexagonal, like those of honey bees, the wasps' approach to this building task is quite different. If a cell has already been started, the soggy mass of wood pulp is applied thickly to the existing walls. The wasps use their heads as the last, building around it. Unlike bees, which work entirely from the inside, the wasps keep one antenna inside the cell and the other on the outside of the wall; they monitor the distance between the two antennal tips to judge thickness. Left unworked, the initial pulpy layer will form a sagging cylinder. Instead, with one mandible on each side of the wall, the wasp works to thin and straighten the structure, squeezing the pulp flat to remove water. In building the cells downward, the wasps take advantage of gravity in drawing out the walls. Cells built sideways or upwards would sag; the worker would have to stay with this alternative structure until it dried.

Just what sort of mental picture, if any, do individual wasps have of their multistory structures? How do they know where new work is needed, and the location of hungry larvae? Wasps are unpleasant to experiment on, and the wages of tampering with the nests of well-armed and short-tempered insects have dissuaded many experimenters. Still, the creatures must employ third-tier mapping to accomplish this impressive task. Moreover, the paths individuals take from the inside to the nest opening at the bottom are, under normal circumstances, anything but random; unless there are unknown gradients of light, temperature, humidity, or odors, the workers must know something about the layout and where they are at any given moment. And in foraging for prey, they need some kind of fourth-tier cognitive map to guide them home.

In the tropics, where insulation is less important, highly social polybiine wasps often do not take as much trouble with the outside covering of the nest. Some still use paper for both comb and container, but

Mapping in Animals

Tier 0: No spatial representation; independent S→R wiring for stimuli.
Tier 1: Internal map: spatial representation of stimuli impinging on body; typically tactile.
Tier 2: Surround map: spatial representation of objects and surface immediately around animal (within one body length, typically mapped by touching); generally tactile.
Tier 3: Local-area map: spatial representation of local objects not within one body length, allowing local navigation through interpolation and pattern matching; typically visual, tactile, olfactory, or auditory.
Tier 4: Cognitive map: spatial representation of the relative position of widely spaced objects or other landmarks, allowing home-range or nest-interior navigation based on a cognitive map; typically visual or tactile.
Tier 5: Network mapping: multidimensional representation of space, tool and/or building equipment, goals, and behavioral options; potential for innovation.
Tier 6: Concept mapping: abstract reasoning, concept formation, potential for insight and language.

construct a rough shell from a single layer of a substance known as *carton*. Carton is a crude, thick paper-and-mud mix not unlike in feel to the substance used in old-fashioned egg containers (which omitted the mud).

The builders of these nests create a roof with a single tier of paper cells pointing down, and then enclose the bottom with carton. Some species build a "double-glazed" version that traps a thin layer of air between an inner and outer lamina of carton. The exit hole is either in the center or at the edge. They expand the nest by building a new nest bottom, and then a wider set of cells pointing down on the old bottom. The new nest opening is directly below the old one, at the center or along an edge.

That wasps are able to use saggy, insubstantial carton to expand their homes at all is remarkable, and certain irregularities can creep into the project. The new bottom, which is built from the periphery toward the center, may not meet at the same height in the middle, or it may come together evenly but leave too little room for cells in the center. These imperfections seem to argue for an absence of de-

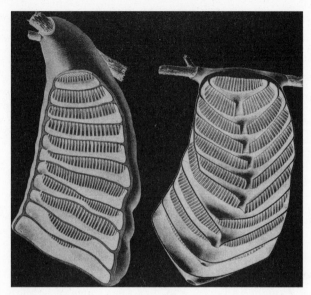

Carton wasp nest. These are nests of two species of polybiine wasps from the New World tropics. Both are built out of carton; the wasps add tiers at the bottom by enclosing the former floor and then extending cells from it into the new layer.

tailed programming and thus, perhaps, a small role for error-prone but more inherently challenging planning and guesswork in the construction. A major difference between paper and carton as walls is that, though repair is easier, the renovation of the hard mud-based carton is out of the question; expansion is necessarily by building additions at the bottom.

A few social wasps have substituted clay for paper. Clay provides more protection against predators and rain. Some small tropical nests have an outer sphere of sand and clay complete with a neat round opening near the bottom. Inside, the wasps use fine clay to sculpt horizontal tiers of downward-pointing hexagonal cells that are identical, except in material, to the single-tier comb of Polistes. The walls are every bit as thin. And on a grander scale, multi-tiered forms are built in the same way as the carton nests, layer by layer.

As in carton nests, expansion is via addition rather than renovation. These can weigh in at three pounds or more after a few years' work. No one knows how these insects recognize good structural clay or deal with variations in quality when building. How they can accommodate the increased stress on the system as the nest grows larger and heavier remains a mystery as well.

ANTS

Except for a brief period in the lives of reproductives, ants are wingless wasps. Nearly all are subterranean, a habitat that not only does not require wings but also actively selects against delicate and easily tangled appendages. By residing underground, ants can survive periods of drought and cold, and thus their range of habitats is very large. This one simple innovation led to a radiation that produced thousands of species, some of which have secondarily returned to terrestrial, or even arboreal, nesting. Given how hard it is to study building in a species that lives under the earth, perhaps it's no surprise that the exceptions, the species that eschew tunneling in the soil, are the best known.

Tropical army ants—both the legionary ants of the Americas and the driver ants of Africa—live in bivouacs. These encampments are constructed of the insects themselves: A layer of ants firmly grasps the underside of a log or branch, then another layer hangs onto these, and so on until sheet after interwoven sheet reaches to the ground. In this matrix of perhaps a million interconnected individuals the ants create nurseries for eggs, larvae, and pupae. There is also an area for the gigantic queen, who spends her days and nights laying an almost continuous stream of eggs.

Raiding columns tens of thousands strong set off daily to forage for this huge colony. As they sweep over the terrain, the strategy of building structures out of colony members can be seen in miniature. For instance, when the leaders of the advancing line encounter a gap in the surface, the first to step into the void freezes with its rear legs holding on to whatever it was last able to walk on. Others to the right and left do the same. The next in line walk across their sisters and freeze in place, holding on to the first tier of ants. And so it goes until a living bridge spans the opening.

When raiding army ants encounter something living, they swarm over it, tear at it, and carry the pieces back to the nest. Often several must coöperate to transport, say, the tail of a scorpion; they have even been seen acting in concert to roll a small bird's egg. The enormous success of their attacks is reflected in the evolution of several species of ant birds, whose diet consists of insects that take flight as the unstoppable column approaches. Yet these devastating killing machines are blind. What forces impel them, and what combination of drive and experience enables their behavior? Can their coöperation, which appears both flexible and rigid, be a product of simple innate instructions? There are hints of possible answers in the group building behavior of the smaller, more experimentally tractable colonies of weaver ant.

Very few ants live in trees. One kind builds carton nests lined with silk for strength and protection from the damp. But the most impressive arboreal ants are the two species of weavers. Since they live in structures constructed in the forest canopy, they have easy access to a high niche out of reach of conventional tunneling ants. Their homes are built as interconnecting chambers of which the floors, walls, and ceilings are leaves still attached to the tree. The leaves continue to photosynthesize, in the process consuming the carbon dioxide produced by the ants and replacing it with fresh oxygen.

Each species of weaver faces two problems: how to draw the leaves together to create a habitable cavity, and then how to attach

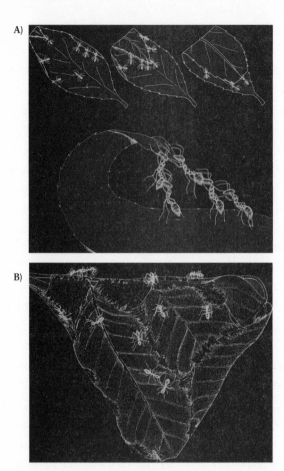

Weaver ant strategy. (A) Ants pulling back on separate parts of a leaf are able to recognize where progress is being made and so redirect their efforts to that spot. (B) Once a leaf has been pulled close to another structure—a second leaf, or part of the same leaf—larvae are used to weave the seams shut with silk.

the edges. The way the ants fold the leaves provides a model for several other examples of group work, perhaps even the bridge-building behavior of army ants. Each insect individually pulls on a leaf edge; this may involve trying to roll back a particular leaf, or standing on one and grabbing the edge of a nearby leaf. The ants may be standing side by side at this point, but most researchers believe that groupings are accidental.

When one ant, or a group of individuals that happen to be pulling in the same direction, achieves some measure of success (a leaf edge begins to roll back, or another leaf is drawn a bit closer), nearby colony mates abandon their own less promising efforts and join the part of the project showing some progress. Their success recruits additional weaver ants from farther away until there is a solid, tightly packed line of workers all pulling together.

This instance of coöperative building, then, depends on the ability of independent insects, each with its own agenda, to recognize a partially completed project, one further along than their own, and join in. In some sense, this is what social wasps do when they encounter an unfinished edge and take up comb or wall building.

The weaver ants employ a unique form of coöperation in joining leaves together to form walls. They bring larvae from the nursery to the work area and move them back and forth; these seemingly helpless immatures in turn exude sticky strands of silk, forming a tight zigzag seam. The living shuttlecocks glue the leaves together, generating an airy but watertight set of joined cavities high above the forest floor.

Many students of insect behavior think that the entire construction process is wired in—that innately recognized stimuli trigger the redeployment of ants to points of progress, and abutted leaves cause some of the weavers to grab larvae and sew. But no one can point to any one cue that would recruit ants from a distance to lend a hand when needed to roll or fold or sew. If these were chimps instead of ants building nests, we would probably assume that a more complex process of multifactor integration must be at work. Surely weaver ants must have some sort of Tier–3 understanding of the layout of their complex, many-chambered nest, and perhaps a cognitive map of the area they forage through as well.

The answers to questions such as these, and the real nature of social intelligence in insects, are to be found in the building behavior and social-control strategies of termites and eusocial bees, a group to which we can now turn.

CHAPTER 5

Bees and Termites

ALTHOUGH SOME ANTS erect haphazard sheets of carton, most build simple straightforward burrows in the soil or in wood. Many kinds of ants erect specialized compartments, but engage in none of the finishing work that might indicate care and planning. Social wasps leave the opposite impression: there is a neat and unitary plan, which is scaled up as necessary. In eusocial wasps the space, whether inside or exposed, has its own independent walls to seal it off from the rest of the world. Coöperation is evident, but there is only a small range of building tasks to be undertaken, only a limited number of placement decisions to be faced.

The nests of bees and termites, on the other hand, give the impression of structures built by a society with a capacity for design skill. There may be many cell or compartment types and geometries, multiple dwelling areas, and elaborate provision for ventilation; in short, the burdens placed on social intelligence and colony control are much greater. We will discover among these two groups the full range of invertebrate coördination and apparent cognition, tactics and intellectual wherewithal that seem well beyond the capacity of most vertebrates. Of course, to the extent that niche—and in particular the aspects that depend on architecture and sociality—selects for intellectual ability, this inversion of our phylogenetic expectations should not surprise us. Social insects need to be smarter than solitary, slow-moving tree sloths.

STINGLESS BEES

Eusociality, the state of communal living so entrenched that individuals cannot survive outside it, arose several times from solitary hymenopterans. One such event gave rise to ants. The vegetarian diet of bees evolved before the change from a solitary lifestyle took place. True sociality has evolved in bees independently many times since, and each time novel strategies of communal nest building, social hierarchy, and communication had to be invented. Thus we might expect to see more diversity in social building and the associated cognitive solutions. In the far more numerous wasps, on the other hand, sociality arose fewer times.

One intriguing development in bee construction is the use of waterproofing material. Some species collect plant resins and tree sap; they mix these with pulp or mud to create a substance called *batumen*, which they use as a protective or insulating case or a waterproof lining. Many species also make wax in special glands unique to bees; this substance may be added to batumen to achieve even better results, particularly at higher temperatures. Beeswax does not sag until well over 100°F, and it is kept carefully below this level by evaporative cooling. Other species use the wax as a thin layer of waterproof varnish around tunnels and cells. Building the cells themselves out of wax was a logical progression, and this independently invented sequence of material use can be seen in nearly any group of bees with both solitary and social representatives.

Groups of social bees living in small colonies include the inconspicuous ground-nesting sweat bees, the less well known allodapine bees (which often make their homes in plant stems), and the roughly two hundred species of bumble bee. The record size for one of these colonies is twenty-two hundred, but most never top a hundred workers. Colonies of stingless bees and honey bees, usually perennial, are generally larger and more intricately designed.

Stingless bees may lack functional stings, but they bite ferociously and eject a burning liquid onto the punctures. So effective is this strategy, and so aggressively is it pursued in the face of danger, that they are the one kind of insect that can defend themselves against army ants. They typically live in hollow trees or burrows in the ground; in either habitat, they maintain only a small readily defended entrance hole.

Colony sizes range from about five hundred to eighty thousand across the two hundred known species of stingless bees. Their architecture is quite varied in general layout, but remarkably similar in detail and function. All use either batumen or a mixture of plant resin, sap, and wax known as *propolis,* which is then combined with beeswax to create *cerumen.* Stingless bees typically employ batumen in defensive structures—layers of armor one to three inches thick to prevent incursions from above or below in a hollow trunk, for instance. Cerumen, on the other hand, is worked into insulating multilayer sheets that look almost exactly like the paper shells around hornets' nests.

All stingless bees build storage pots for honey. Honey is a remarkable substance, and since like wax it is made only by bees, the steps leading to this feat of chemical engineering probably evolved only once; very likely it played a key role in the subsequent speciation of this group. The characteristic of honey essential to colony survival is that it can be stored, an innovation that helps to insulate bees from fluctuations in the current nectar market. It is made from ordinary nectar treated with an enzyme that alters a bond between the sugars, making a liquid indigestible to yeast and other fungi. The bees then thicken it by fanning air over the chambers. In time the honey becomes so viscous that bacteria become stuck and die in their own microscopic envelope of toxic wastes. Just before their

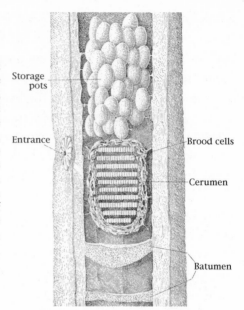

Stingless bee nest. Although these nests superficially resemble those of some eusocial wasps, the brood cells open upward and are built of wax. The insulation consists of cerumen, and is anchored on the sides to the enclosing trunk. Storage pots for honey are unique to bees. The layers of batumen, which seal off the nest from the rest of the hollow trunk, are another design feature not seen in wasps.

Storage pots

Entrance

Brood cells

Cerumen

Batumen

stored honey is ready to crystallize, the bees cap it to keep it liquid. When too little nectar comes into the hive to fuel the workers, the bees sip a bit of honey.

Stingless bees build their ovoid honey pots out of cerumen, as well as the pots they use to store pollen, which serves as the hive's protein source. To keep pollen from spoiling in the humid and warm interior of the hive, the bees must treat it with a fungicide. In some species, the pollen pots have a distinctive inverted-cone form. Ovoid or spherical, these storage vessels are much larger than the cells for rearing brood. Larvae growing in the brood cells are fed with a mixture of pollen, honey, and water.

Wasps, being carnivorous, do not need pollen and honey chambers, but we've already seen that their principle of erecting a multilayer insulating sphere turns up in some stingless bees. Another feature that these bees share with wasps is the horizontal arrangement of tiers of brood cells. At first glance the similarity is striking, but a closer look reveals important differences. The bees' cells are made of cerumen rather than paper; since the larvae do not stick to the waxy building material, the cells open up rather than down, and their support columns are less regularly placed.

A few species do without perfectly horizontal combs. Instead, brood cells are built independently, suspended in a free-form design by columns attached to the wall, other chambers, or pollen pots. Though functional, the result fails to exploit the economy in material enjoyed when cells are built sharing walls. This economy is important since beeswax is metabolically expensive to make: fourteen ounces of sugar (equivalent to about seventy-five ounces—four and a half pounds—of nectar) are needed to synthesize one ounce of wax. And the higher the proportion of beeswax in the cerumen, the stronger and more resistant to sag the finished product is.

Whatever the arrangement of the brood cells among species, they are all single rooms. Bumble bees, on the other hand, raise several larvae in a large cell, which minimizes the ratio of wax to larvae as effectively as shared-side comb cells. Disease probably spreads faster in multilarvae nurseries, however, and there is always the very real risk of cannibalism; but the single cells of comb exact a long-term cost as well. Each larva spins a cocoon before pupating; when it emerges, some or all of the silk cocoon is left behind, incrementally reducing the interior diameter of each cell. Bees reared in comb that has housed generations of larvae simply cannot grow as large before pupating. Some of the workers in older colonies, the ones who were larvae in longstanding hand-me-down comb, will be noticeably smaller.

Stingless bees, like honey bees, employ a remarkable division of labor. What is accomplished in ants and termites through the creation of separate castes of workers is done with perhaps less efficiency but more flexibility in highly social bees; the newly emerged bee begins a seemingly regimented, ever-changing series of tasks that ends in foraging. The youngest bees typically clean cells most of the time. This activity segues into feeding the queen and the larvae. As their wax glands mature, workers spend more time building. The next job they take on is to unload returning foragers and stow their pollen and nectar. Then workers begin the task of guarding the hive entrance, and soon they are foraging, a risky activity normally continued until they die of old age after three to five weeks of constant commuting.

Since the castes of ants and termites that deal with the larvae are the smallest of all the workers, it stands to reason that small bees might be better at this task. And since guard ants are invariably the largest caste, having a special size group for this job may make sense. But whatever inefficiency is created by the honey bees' one-size-fits-all strategy must be weighed against a problem that arises in colonies with castes: if a sudden attrition of guard ants occurs after a battle, this security force can be replaced only after the queen lays new guard-ant eggs, the larvae hatch out and are reared, and the pupae metamorphose and emerge. This takes from two to three weeks. In bee colonies, a loss of guards causes those in proximal occupations—the younger unloaders and older foragers—to accelerate their progression or revert back to take up the slack. The same holds true for nurse bees and builders.

The cognitive implications of a jack-of-all-trades approach are substantial. A bee must be able to take on any of a number of complex tasks, whether it be foraging (which requires solar navigation, use of cognitive maps, and an ability to memorize a plethora of cues associated with nectar-producing flowers), or building (which could require fabricating pollen pots, or honey cells, or brood comb and its supports, or mixing batumen). In addition, these job descriptions are not as hard and fast as they might sound; nurse bees will also clean a cell if the larvae are in good shape, or do a little cell trimming if an unfinished comb edge presents itself. Bees of any age will pitch in when needed. This is a clear instance of the second level of social task mapping.

Social Intelligence in Animals

Social 0: Social isolation; conspecifics indiscriminately either ignored or attacked.

Social 1: Social hierarchy: linear representation of part or all of the social order, especially individuals ranked near the individual in question.

Social 2: Decision-network mapping: multidimensional representation of parameters important in making social choices.

Highly social bees are also unusual among social insects—indeed, among animals in general—in the way they invest in offspring. Ants, wasps, and termites periodically produce a huge wave of reproductives that set off to mate and begin new colonies. Since the world is not overrun with these insects—at least not until picnics in late summer—the success rate of the offspring is clearly very low. Lots of animals pursue the approach of many small, "cheap" offspring; biologists call it *r*-selection (a reference to the characteristic reproductive rate of a species). The other strategy, *K*-selection, is controlled by the number of organisms the habitat can accommodate. Animals and plants that reproduce by this system produce few offspring, but they lavish time and metabolic resources on them. This approach is seen in stingless bees and honey bees—and in humans. A reproductive unit in itself, a bee hive sends about half of its members out to establish a new colony. Stingless bees provision a new nest with cerumen taken from the old hive, along with honey and pollen. Construction of a full-scale hive begins at once, undertaken in large part by bees that have technically passed through the building phase.

As we have already mentioned, bees work the cell walls somewhat differently from wasps, a difference that makes good sense considering the nature of wax and cerumen versus paper. Recall that wasps work exclusively at the growing end of a cell, squeezing out the excess water by compressing the wall from each side with their mandibles; they judge the thickness by the separation between the tips of the two antennae (one in each cell). Bees, on the other hand, chew the wax mixture until it is soft, then spread it on the growing edge of the cell. The bee puts its head in the cell; the mandibles are used to sculpt away excess wax to be used later.

Antennae are used in conjunction with the mandibles to achieve a perfect degree of flatness and uniform thickness. With the mandibles just touching the wall, the antennae reach out to the surface slightly to each side of the mouthparts; if there is a bulge under the mandibles, the excess is scraped away. If not, the mandibles gently shove the wall, and the antennae judge whether the elastic wax gives just the right amount. All this work is done in the dark. And as with the wasps, the diameter of the insect's own head is the form used to determine the diameter of the cell.

The cognitive implications of stingless bee architecture are hard to judge on their own. They are sufficiently difficult to study that even the basic life cycles and nest structures are known in only a few cases. E. O. Wilson concludes that they have evolved behavior as complex and flexible as that of the well-known temperate-zone honey bee. For example, stingless bees share information about food. This potential advantage of sociality is of enormous cognitive importance. Although it has seldom evolved in nature, information exchange about food reveals a flexibility that goes beyond programming. Here is how it works: a returning forager who has found a rich source of nectar or pollen will run about the hive buzzing. Unemployed foragers, who smell the odor of the flower adhering to the waxy hairs of her body, become excited. In some species, this is almost the end of the story. Excited foragers set off and search for the right odor until they find the food source that the successful forager has visited.

At the time this phenomenon was first observed, researchers assumed the recruited bees hunted at random. But we know now that honey bees create (among other things) a fourth-tier olfactory map of their home range, one of many cognitive abilities that humans lack. Smelling a familiar odor on a buzzing forager, they set off for where they remember having encountered it previously. And since honey bees can detect and recognize the distinctive odors of different locales, it's reasonable to assume that stingless bee recruits can also begin a nonrandom search without having previously come across the particular species of flower being advertised, and a certain amount of behavioral evidence supports this guess.

For other species of stingless bee, finding the target is much easier. The forager makes her return flight by means of short steps, leaving special scent marks every two or three yards along the way. After alerting recruits, who assemble near the hive entrance, she begins the tedious process of leading them along her odor trail from one olfactory mark to another. Eventually, some of them reach the food, and can then navigate back to the hive on their own using their ability to integrate the separate outward legs of the journey; olfactory information is irrelevant after finding the goal. This system of guidance is fairly efficient over a few dozen yards, and where there is a relatively flat surface suitable for marking. For longer distances and through scrub, the olfactory clues are hard to find. Nevertheless, few vertebrates can manage anything so impressive.

We can infer something of the mental processes behind the building behavior of highly social tropical bees by putting ourselves in their position, envisioning life inside dark, elaborately constructed three-dimensional nests while also foraging in the complex three-dimensional world outside. For example, we can ask whether the behavior of the workers shows that they have only local Tier–3 knowledge of the nest—which general area they are in, and the repeating structure of the region—or a Tier–4 cognitive map, which provides a representation of the entire structure. Is their ability to navigate through their forested tropical habitat based on regional (Tier–3) reactions to local cues (odors, for instance), or does it depend on a true Tier–4 representation of the home range? Most observations suggest the higher level of mapping ability, but these small tropical insects are very hard to experiment on, and their habitat adds to the challenge. To date, we can only infer the cognitive capacities of stingless bees. Researchers have discovered far, far more by looking instead at the repertoire of their more easily studied counterparts, the hive bees.

HONEY BEES

Honey bees, like stingless bees, evolved in the tropics. The earliest species were probably very much like the dwarf honey bee *(Apis florea)* of India, Southeast Asia, and the East Indies. These tiny insects live on a single bare honeycomb in colonies of several thousand. Like all honey bees, they differ from stingless bees in that their combs of hexagonal cells hang down vertically. Cells are built out horizontally, attached back to back in a double layer. The center of each cell on both sides is at the junction of three walls on the other, a structural trick discovered millions of years later by human engineers. The sheets of comb are constructed entirely out of beeswax, and they are so strong that one ounce of wax honeycomb can hold about two pounds of honey, pollen, larvae, and pupae. (In a few colonies near volcanoes, where the air is exceedingly hot, the bees mix into their wax an extremely stiff resin to raise the melting point; whether this behavior is part of an innate but obscure contingency plan, or is somehow learned by the bees, is unknown.) Another difference compared to stingless bees is that the cells for honey and pollen are identical to the ones used for brood; whether this is more efficient and flexible or quite the opposite is not obvious.

The comb of the dwarf honey bee is built up and around a branch and is entirely exposed, though the wax over the branch widens to provide a kind of awning on each side. The colony depends on coöperative guarding for its survival. Since at least half the bees in a nest are not yet foragers, the supply of defenders is large; but the nests, hidden as they are among the leaves of a tree or bush, are fairly cryptic. Few are discovered by predators, but those that *are*

 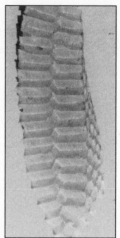

Honeycomb. (A) From face on, the precise hexagonal arrangement and exceedingly thin cell walls are evident in this small section of comb (which has about a hundred cells visible and another hundred on the back). An average colony will have several sheets of parallel comb with about five thousand cells in each. (B) In cross section, the elegant way in which the bases of the cells are staggered to increase strength is clear.

found provide the most concentrated natural source of carbohydrates on earth. It's hard to imagine the days before commercial sugar—cane sugar, corn syrup, and beet sugar—when the sweet cravings of humans and other animals led them to seasonal fruits and, when the opportunity presented itself and the risk seemed worth the taking, the intense sugary honey of bees. The main reason we know so much about the four species of *Apis* is that for centuries they were the one source of sweetness that could be stored indefinitely. The desirability of beeswax is the rest of the story.

The preservative powers of honey are legendary. When Alexander the Great was killed in his last campaign, his body was placed in a gold coffin filled with honey for transport home and burial. Given the high cost of honey, it's not obvious whether the preservative or the ornate coffin was the more outrageous example of conspicuous consumption. This kind of outlay on dead rulers is not limited to the Greeks and the pharaohs of Egypt: in England, the first four earls of Southampton were also buried in honey-filled caskets—honey that was still free flowing and sweet, according to the workmen who excavated the containers and sampled the contents centuries later—before discovering what lay inside.

Communication

Like stingless bees, dwarf honey bees have an information-sharing communication system, but the honey bee system is infinitely more sophisticated. The fascinating dance language of bees speaks volumes about the cognitive potential of insects in general, their communication, decision making, and cognitive maps, all critical components of building behavior. It is a window into their minds and it shows us how some things work, and how others might operate. Although the dance language was first decoded in temperate-zone honey bees, it evolved in a more comprehensible form in the dwarf species first.

Successful foragers from the dwarf honey bee hive return to a platform over the branch that supports the hive, where recruits are waiting. There they perform buzzing runs on the horizontal surface much like those of stingless bees, except that the dances are performed on exposed comb, not inside a dark cavity, and they point directly at the food source. The foragers use the sun and other celestial landmarks to orient their dancing, and recruits employ the same cues to decode the message. In addition, the duration of the buzz is directly related to the distance from the hive to the food. While the forager runs, she waggles her abdomen back and forth, and at the end of a run she hurries back along a semicircular route to the starting point to begin the waggling buzz again. She also alternates left-handed returns with right-handed ones; attending bees seem to expect this, and anticipate her movements. Recruits are able to interpret her message and can fly to the vicinity of the food. Although the recruits can smell the odor of the food (and, presumably, its locale) on the bodies of the dancers during the waggling runs, for most conditions the dance language point-and-buzz system is a better way to recruit quickly and accurately.

The familiar temperate-zone honey bee, *Apis mellifera,* derives from the phylogenetic descendent of dwarf bees, the Indian honey bee *Apis cerana.* Like the stingless bees, *cerana* has discovered the defensive advantages of living in a hollow tree. Brought to the New World by Columbus and other travelers, honey bees have been able to survive by living in insulated nests in which they construct several sheets of vertical comb, spaced two bee diameters apart. The dances cannot point toward the food because there is no horizontal surface for dancing. Moreover, it is dark inside the nest, so there are no celestial cues for orientation. The bees have solved this problem in a way that baffled observers until 1945, when Karl von Frisch finally saw how it works. The dance retains the figure–8 form of buzzing waggle runs and semicircular returns, but the orientation of the dances on the honey bees' vertical comb is referenced to gravity. No longer do the dances point directly at the food; "up" is taken as the direction of the sun, and the dance points to the left or right of vertical by just the angle between the food and the sun outside. So, for instance, if it is solar noon and the sun is due south, and the food is also to the south, the dance will be oriented straight up on the comb. If the food is in the east, the dance will point 90 degrees to the left of vertical; west is indicated by waggles 90 degrees to the right of up. As the sun moves from the east to the west, dances to a particular site precess counterclockwise in compensation.

The dance communication system is called a language because it uses arbitrary conventions to describe objects or events distant in space and time: The food lies out of sight hundreds of yards away, having been visited up to half an hour previously. All honey bees agree that *up* is the direction of the sun, but it could just as easily have been the direction the hive entrance faces, or north, or even

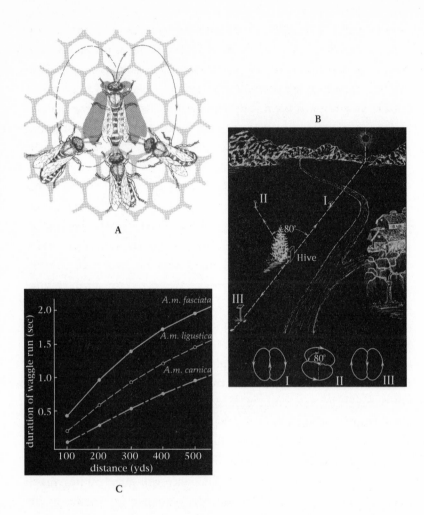

Honey bee dance. (A) The dance is in the form of a figure–8, with the dancer waggling from side to side and buzzing during the straight central portion of the manoeuver. (B) The direction of the dance on the vertical comb in the dark interior of the hive encodes the direction to the food outside. If the food is in the direction of the sun (I), the dance is aimed up; if, instead, the food lies 80 degrees to the left of the sun (II), the dance points 80 degrees to the left of vertical. (C) Distance to the food is encoded as the duration of the waggle run, the number of waggles, and/or the number of sound bursts (all of which are tightly correlated). The actual conversion depends on the subspecies of bee: Egyptian bees produce more waggles for any given distance than German bees, the Italian dialect falling in between.

straight down. The exact convention does not matter, so long as all members of the society agree on its use. This point is underscored by the subsequent discovery of a dozen different conventions in coding for distance in the temperate-zone honey bee. To the Egyptian subspecies of honey bees, a waggle means about ten yards; to the Italian subspecies, closer to twenty; and to the German variety, forty. These dialects are innate; pupae from one subspecies that emerge in the colony of another dance according to the conventions of their genetic, rather than their adoptive, sisters.

The dance language makes it possible to observe foragers in a small glass-walled colony informing their hive mates where they have been. Manipulations of the foragers' journey tell us that they understand a great deal about their world. They can easily be trained to an artificial food source by offering sugar solution in a feeder and moving it in short stages from the hive to an experimental location; foragers frequenting the feeding station can be marked with paint dots, and their dances observed back in the hive.

For instance, Karl von Frisch once trained forager bees around a long thirteen-story building. Once there, von Frisch increased the sugar content of the food to make it more alluring; the foragers, who flew to and from the feeder along a route around the end of the building, began to dance upon their return. The dance indicated the distance and direction of a site they had never flown to directly, as if the building were not there. For a seventy-five-yard flight to a target that was actually thirty yards away, they got the distance right to within four yards, and the angle was a mere five degrees off. To achieve this degree of accuracy, humans would have to resort to maps and trigonometric calculators.

Maps

Bees exhibit so much knowledge about their surroundings that by 1980 researchers wondered whether bees might have some sort of

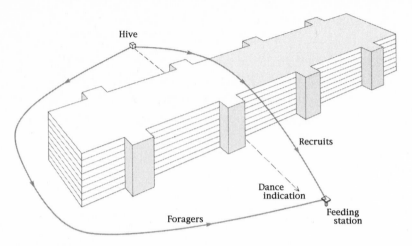

Detour experiment. Foragers trained to fly an extreme dogleg route around an eight-story building to a feeding station were nevertheless able to indicate by their dances a location very close to the actual site. Recruits stimulated by the dancing flew the true direction by going up and over the building.

cognitive map. Just imagining this was a stretch; psychologists had only recently and disconcertingly discovered that rats have cognitive maps, and the idea that a half-inch insect might as well was difficult to believe. To test this, experimenters trained a group of foragers to one location (A), out of sight of the hive, and then after letting them feed there for a couple of days, captured them as they left the hive en route to the food source. The captured bees were then taken to another location (B) out of sight of the food source but well within their home range, and released. The foragers circled up and departed directly for the unseen feeding station from this unexpected location.

Not surprisingly, this experiment created a great deal of controversy. Insects were assumed to do everything in the simplest possible way, and the idea that they had any sort of map sense was unsettling. For a time, subsequent experiments meant to investigate the phenomenon consistently failed to take into account the limita-

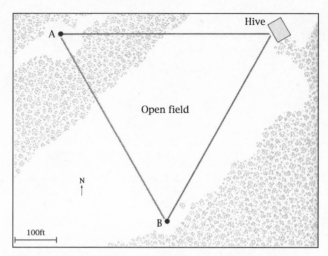

Cognitive-map experiment. Foragers were trained to feeding station A in a forest clearing. After visiting this feeder regularly, the foragers were captured individually as they emerged from the hive en route to feeder A. These kidnapped foragers were carried in a dark container to site B and released. Their vanishing bearings demonstrated that they were able to orient themselves and fly directly to site A.

tions the bees work under, such as extreme nearsightedness, and the need for an opportunity to familiarize themselves with the landmarks around their hive in order to generate a map. As a result, it took fifteen years to parse out the bees' unique talents. The prospect of invertebrate intelligence has been a difficult concept for humans to grasp, and the more we learn about honey bees, the more awesome their accomplishments appear.

For one thing, foragers can recognize rotated images: trained to one pattern, they can choose a rotated version of it in preference to an otherwise similar pattern. Human intelligence tests once included mental figure rotation as a test of cognitive power, and the ability to manipulate mental images has long been seen as an indication of the ability to form concepts.

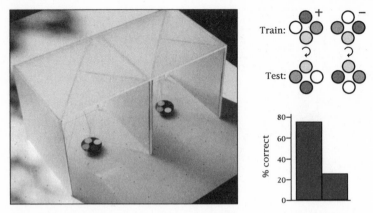

Mental rotation. (A) Foragers were trained to choose the correct feeder in an apparatus that restricted their angle of view to a range of 90 degrees. The feeder with nectar and the alternative offering only water were switched between sides to prevent the bees from learning a left/right distinction. (B) For testing, new feeders with the same patterns were set out, but the patterns were rotated 180 degrees to an orientation not previously visible. Foragers chose the correct alternative from 75 to 80 percent of the time, indicating that they could rotate their memory of the learned pattern in their brains.

Concept formation involves recognizing the basic abstract qualities that identify a stimulus, whether visual, acoustic, or olfactory, while ignoring aspects that may vary. In a typical test, for instance, the shape, color, and size of a target image will be changed between visits or trials, but the left/right symmetry of a rewarded visual cue will remain consistent. When tested, bees can select symmetrical patterns they have never seen before simply because they are symmetrical; if trained to pick asymmetrical targets, they will generalize to patterns sharing that conceptual quality. Essential to our language, and once thought to be uniquely human, concept formation has been found in animals across phylogenetic lines, from pigeons to dolphins, from bees to apes. This souped-up use of mental maps—the mapping of specifics into abstract generalities—corresponds to Tier 6, and may make possible a kind of reason-based problem solving that depends on manipulating concepts.

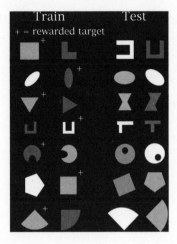

Concept formation. During training on a series of shapes (left pair of columns) individual foragers found food (+) only on targets that displayed horizontal (mirror-image) symmetry. Left/right placement, color (indicated here by shading), and shape had no predictive value. When tested on new pairs of shapes (right two columns), the bees readily chose the novel symmetrical patterns. Training to asymmetry was equally effective.

Mapping in Animals

Tier 0: No spatial representation; independent S→R wiring for stimuli.

Tier 1: Internal map: spatial representation of stimuli impinging on body; typically tactile.

Tier 2: Surround map: spatial representation of objects and surface immediately around animal (within one body length, typically mapped by touching); generally tactile.

Tier 3: Local-area map: spatial representation of local objects not within one body length, allowing local navigation through interpolation and pattern matching; typically visual, tactile, olfactory, or auditory.

Tier 4: Cognitive map: spatial representation of the relative position of widely spaced objects or other landmarks, allowing home-range or nest-interior navigation based on a cognitive map; typically visual or tactile.

Tier 5: Network mapping: multidimensional representation of space, tool and/or building equipment, goals, and behavioral options; potential for innovation.

Tier 6: Concept mapping: abstract reasoning, concept formation, potential for insight and language.

The pioneering animal intelligence researcher Donald Griffin suggested that if a particular cognitive capacity would be useful to a species in its niche, we should look to see whether something has evolved to fill that need. What he may not have realized is that changes in cognitive capacity can alter the nature and breadth of a species' niche. It is with these two possibilities in mind, and the knowledge that selection has already supplied honey bees with the second-most complex language on the planet, a navigational system that lacks only a GPS receiver, the most extensive memory capacity of any insect known (as yet), not to mention maps and concepts, that we will begin our look at their building behavior.

Starting a New Hive

Like stingless bees, honey bees reproduce by sending out about half the members of the old colony to found a new one. But there are important differences. A new stingless bee queen leaves with the swarm, departing for a new home that has already been chosen. The honey bee colony rears a new queen, but the old queen leaves with the swarm, which hangs temporarily from a nearby tree branch while scout bees scour the surrounding countryside for nesting sites. The scouts then dance on the side of the hanging mass of bees to indicate the distance and direction of the cavities they have found.

The process of choosing the location of the new hive is an elaborate program of comparison previously unimagined in insects. After dancing for her site, a scout may return for a more thorough inspection. As the bee paces off the dimensions in the dark, and flies from one corner of the cavity to the other, this inspection can take as much as half an hour. From clever choice tests performed by Tom Seeley and others, we know that the ideal site is well off the ground, has certain minimum and maximum volumes, is taller than it is wide, has an unobstructed south-facing entrance about half an inch wide, is dry and free of drafts, and is a particular distance from the home colony. The bees allocate about four days to this task, then they make a commitment.

Tracking the dances on the swarm reveals a pattern of shifting loyalties and fads. Dozens of cavities at a variety of distances may be advertised. Other scouts visit the competing sites and return to

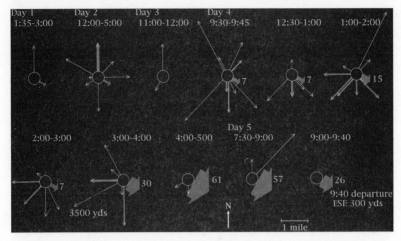

Swarm dancing. Returning scouts danced on the side of the swarm, indicating the location of a suitable cavity. The dancing, followed by inspection visits from other scouts, enabled many bees to compare the prospective nest locations. The thickness of the arrows indicates the relative number of dances to each site. Over the course of five days, a consensus emerged for a nest site three hundred yards away to the east southeast.

dance for the one that seems best, weighing one thing with a dozen others. Even bees that are happy with their own entry in the competition revisit it just to be sure, and they abandon the property if an experimenter pours some water into the cavity. In time a consensus is reached, and the swarm departs to set up housekeeping. Were this description rewritten substituting "ape" for "bee," we would still be amazed at the cognitive flexibility of the behavior.

Once the bees are in the new cavity, there is not a moment to be lost. The swarm needs to build a substantial amount of new comb before the three-week process of generating new workers can even begin. In a few weeks, before the nectar flow begins to decline in late summer, all the honey that will sustain the hive through the winter must be brewed and capped. But the four days of decision making are essential: although 90 percent of established hives survive the winter, fewer than half of the new nests will make it, so the very best place must be found. Getting an academic committee of two to agree on anything is tricky; integrating the independent judgments of scores of scouts from the hive's research and development division seems incredible.

Getting right to work fabricating comb presents an immediate challenge. How can ten thousand builders agree on where to start the parallel sheets of wax, or in what direction to build them? The new cavity is likely to be asymmetrical, or to have unusable areas. A worker with wax and the motivation to build will search about and deposit her flake on the ceiling. Others will be doing the same. When one builder finds the minute beginning of comb at the top of the cavity, she is more likely to add her wax to it than to find another spot. Just as with the weaver ants, any substantial start draws workers to focus their efforts on the cornerstone spot.

From this point, building centers along what the bees picture as the long axis of the horizontal plane of the cavity, and all other bits of wax elsewhere on the ceiling are scavenged and added to the growing sheet. As the first piece of comb is lengthened down from the roof, the base on the ceiling is widened to almost the full width of the space. Soon parallel beginnings start to appear, leaving always two bee diameters between the cell openings of adjacent sheets. If an irregularity in the roof creates a bend in the first comb, that irregularity is faithfully reproduced in the other sheets.

But if the cavity is not considerably wider in one dimension than the other, a potential arises for confusion about the direction the comb should face. The builders, working rapidly in the dark, settle the issue by erecting the new comb along the same axis they grew up with in the old hive. This is another of the useful arbitrary conventions honey bees resort to in aid of efficiency. But how can they judge this angle? Tests in which new hives were surrounded with magnetic coils show that the bees use the geomagnetic angle remembered from days or weeks before to establish the basic orientation of the sheets. Imposing a strong radial field leads to a weird, curving comb with useless intersecting layers, blind ends, and sealed-off chambers.

Organization and Control

It is important that a colony of tens of thousands of independent insects make sensible, coördinated decisions in these massive coöperative building efforts. It is as important that some bees elect *not* to build, but instead to forage, tend the queen, guard the hive, remove dead bees, feed the larvae, and so on. Honey bees have mistakenly been taken as an ideal of centralized socialist control; instead, they are a pretty good illustration of free enterprise, of Adam Smith's Invisible Hand evening out supply and demand. Even modest unemployment is critical for taking up the slack as needs change. And the same every-bee-for-herself system that allocates individuals among the various tasks also regulates the details of any particular one—the shape of comb, the capping of cells, the building of some slightly larger-caliper drone cells, the construction of queen cells prior to swarming, and the modifications of the interior using propolis to fine-tune the size of the opening and waterproof the interior.

The key to such a system is that the colony avoids the extremes of unanimity (what we might call "behavioral correctness") on the one hand and iconoclastic diversity on the other. Watching the way bees build queen cells shows how this balance is achieved. Queen cells appear when the colony is preparing to swarm, or when the queen is dying. The premeditated cells that supply new queens in advance of colony reproduction are built down from the bottom of the comb or, less often, from the ceiling of the cavity. They are tapering cylinders of wax far thicker than ordinary comb; the queen lays an egg in each, and the larvae that hatch are fed royal jelly, a rich glandular secretion made by nurse bees. This substance activates the genes that program the development of a queen, with her small eyes, long body, degenerate brain, and enlarged ovaries.

Most hives show the tentative beginnings of queen cells. About one bee in twenty seems to notice the small start of a cell, and most of those barely break stride. Perhaps one in fifty pauses and explores the bit of comb with her antennae. If the queen is healthy and no swarming is imminent, most of these bees will rip off part or all of the cell beginning and carry it off. A minority, however, will add some wax. Each bee seems to be reading the need for a queen cell based on chemical signals in the hive (mostly the level of queen substance, a potent pheromome produced by the queen and circulated by the workers) as compared to her own personal cell-building threshold. The bees vote with their wax; if more add wax than take it away, the cell is completed, but if more remove wax, no queen chamber ever gets far.

The hive operates in a state of dynamic equilibrium. Just as we achieve delicate finger movements by stimulating extensor muscles and flexors simultaneously, controlling the relative contraction of each so that they pull against one another, the wax removers and the wax adders are pulling against each other's efforts. The result is a graded response, tuned to the needs of the larger entity (the body or the hive), always ready for a quick response. But no central nervous system is available to the bees: each creature acts independently. It is essential that they have different thresholds, or this and most other behaviors would be all-or-none responses, with thousands of workers throwing themselves suddenly into building or guarding or nursing. On the other hand, they need to agree generally that low queen-substance levels mean it's time to build queen cells, though they may differ on how low is really serious.

The different thresholds arise, like most things, from a combination of chance and causation. Chance here is how recently the bee in question encountered another bee bearing the queen's powerful odor. Causation is the bee's general motivation to build, which is compounded of its age (to what degree it is in its building phase) and genetic endowment (some bees are predisposed to build, but others need quite a lot of stimulation to respond to construction cues). Something similar goes on with foragers: All hives have a hard core of scouts, loose cannons who refuse to attend dances and learn from their sisters. Instead, these bees seem driven to find new food on their own and then, after advertising the discovery, are impelled to look for something different. Most forager bees, though, are determined conservatives, never searching for anything not described in an official dance. There are independents, though, who switch parties depending on whether the hive is enjoying a time of plenty or is desperate for new sources. We know these thresholds are at least partly genetic because hives can be bred to either extreme, both of which are fatal to the hive.

This rich, flexible control system, with its multiple levels of operation, is a classic instance of social network mapping, the level at which social choices demand a complex multidimensional awareness (Social 2). Solitary insects have no need for this level of control since there is no need to consider the behavior of other individuals, and no necessity to coördinate responses. At any given time, the linear life of individual build-and-provision insects generally has one overriding priority; social insects face a multitude of choices and problems, the correct solution to which depends on the current and likely future behavior of others. And the apparent ability to factor in the probable short-term decisions of conspecifics, and the social inertia and momentum that go with it, open new cognitive possibilities for increasing yet further the efficiency of the colony.

A good example of behavior organized along these lines is the regulation of hive temperature. When the hive is too hot, bees fan their wings to cool it down; if it's cold, they vibrate their flight muscles to generate body warmth. In a typical outbred colony, different bees have their own individual set points for initiating cooling or heating; their thermostats can vary by as much as five degrees, leading to a seemingly inefficient graded response to the air conditioning needs. But in inbred colonies, the workers may all agree on what is too cold and when it's too hot, and a sudden universal move to evaporate water or shiver the flight muscles can lead to temperature variations three times the size seen in outbred colonies. The multiple-threshold approach is a sophisticated kind of negative feedback system; bees act to correct variations from a target range, but do so without overreacting.

Honey bee dancing exhibits the same principle of dynamic control. Most foragers do not dance when they return with food. Even bees visiting a spectacular site do not all advertise it, and an occasional forager will reel off a few cycles about a poor source. Again, chance and causation are both at work. One of the biggest factors is how long a returning forager goes before a hive bee relieves her, proboscis to proboscis, of her load; if it takes less than forty-five seconds, she is more likely than not to dance. Unloading depends on how good the food is (unloader bees may refuse to take low-quality nectar from another worker) and whether even a forager with excellent wares chances to find an empty and willing taker.

Consider an experimental feeding station offering a mildly sweet sugar solution, that elicits dancing from about a quarter of its visitors. At 1:00 P.M., another group of foragers visiting a second station at the same distance but a different angle discovers that the weak brew they have been dutifully collecting has been replaced by quite sweet sugar water. When bees return with the potent food, they catch the attention of many unloaders, and so a high proportion of these foragers begin dancing. Within forty-five minutes, foragers from the first station will have stopped dancing altogether, even though their food is still the same. At 2:30 P.M., the good times end at the second station, and those bees go back to collecting their former thin gruel. Now the first-station foragers find a warmer welcome in the hive, and are back to their normal unenthusiastic dancing within half an hour. Independent judgments by each forager and each unloader, along with free-market competition, regulate the dancing.

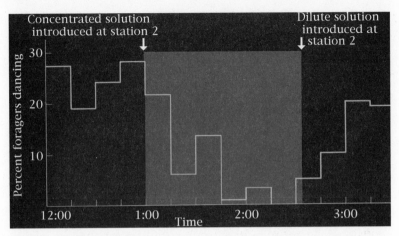

Station competition. Dances to one station offering the same sugar solution throughout the three and a half hours of the test are plotted. At 1:00 P.M., an hour into the experiment, a second station began presenting more concentrated sugar water, and its foragers started to dance vigorously. During the hour and a half that followed, dancing to the first station fell off dramatically on the basis of feedback from dance attenders in the hive. When the food at the second station was replaced with a very dilute solution, dancing to the first station recovered to roughly normal levels within a half hour.

The foragers factor in other variables as well. Most obviously, pollen collectors would never dance if audience feedback were all there was to it. Hives need pollen as a protein source, but pollen gatherers have to unload themselves into empty cells. Even nectar foragers have important cost-benefit information not available to the bees in the hive: how far away the food is, how difficult or time consuming it is to extract from the flowers, and so on. Each bit of information changes the probability of dancing, but in a somewhat different way for individual foragers; some factors that might be called aesthetic components (that is, economically foolish factors) seem to be weighed in the balance. Foragers like close-fitting flowers that they have to push into, narrow tubes holding the nectar (though these are harder to empty), blossoms with less (but not zero) odor, flowers with lots of petals, especially if they are violet or blue, and so on. And chance affects their judgment—how many of the first few flowers the forager encounters are full of nectar, or have instead been recently emptied by another bee. Even the weather matters; collecting continues, but dancing stops for more distant sources when a storm threatens. Note, too, the need here for a certain percentage of unemployed workers. Without idle foragers, the colony could not move quickly to exploit a promising new source.

Could the phenomenon of dynamic equilibrium account for the adaptive variability and quick response to need in a hive of twenty thousand basically identical robots? The decisions of foragers, blending as they do multiple factors both practical and aesthetic, read like elaborate and rather stuffy restaurant reviews: the quality of the entrée, its "presentation," the ambience, location, price and distance, aromas and tastes, not to mention the competition. But although we tend to ascribe a fine degree of conscious discrimination to food critics, honey bees seem at first glance less likely to be aware of their decision-making process. But couldn't we program a robot to judge human food? We could even build in enough random noise or sampling error so that different robotic critics would disagree, and award different numbers of stars—or probabilities of dancing, if that is the way you score the establishment.

Or perhaps we are doing just what the bees refuse to do: committing ourselves entirely to one side or the other. Perhaps building, foraging, and dancing are optimized mixtures of automatic and planned behavior. In judging food, we use an enormous array of prewired sensory and neural processing circuits—olfactory, gustatory, tactile, and visual. Our reactions to what we eat and drink blend the innate and the learned, the "sensible" and the aesthetic. When we consider how much bees can learn about their hive and surroundings, and how much they are instinctively driven to learn about them, and if we take into account their ability to conceptualize and categorize, the facility with which they form and use cognitive maps, and their finesse at weighing a dozen factors with at least as much accuracy as we do, it seems foolish to relegate their construction ability to rote programming.

Honey bees, then, enjoy a rich innate repertoire of techniques. No individual bee, for instance, has time to learn how to work wax into hexagons of precisely the correct thickness, and to hone her skills to the level of mastery the hive requires. These innate abilities could be best put to use if bees can picture where they are and at what stage the building process is, and how they can best advance the project. This third-tier behavior easily fits a hybrid strategy of using a wise blend of instinct, learning, and cognition to orchestrate behavior. Given the readiness of selection to seize upon opportunity and capacity (and honey bees have a surfeit of capabilities), and the life-or-death importance of the tasks, it would be surprising if evolution had opted for a mindless automaton approach to nest building in honey bees.

Positive Feedback and Evolution

We have been careful to emphasize how social control within an insect colony usually depends on elaborate negative feedback. There is a dynamic balance that relies on the independent decisions of thousands of individuals, and this balance almost always prevents the group from doing anything too rash. Negative feedback is a nearly universal characteristic of biological systems. When our internal sensors find that we are cold, they trigger shivering and higher basal metabolic rates to warm us. Once the body reaches the right temperature, these cellular thermostats turn the heating off. Similarly, when we become too hot the sensors trigger sweating, only to cancel that order when the body temperature returns to its target range. Circulating hormone levels, heart rate, blood pressure, and muscle movements are controlled in the same self-limiting way.

But sometimes positive feedback is called into play. This is basically a biological amplifier, and the response itself triggers additional responses. We are most aware of positive feedback when it misfires, as in an allergic reaction or a panic attack. But inside cells, the proper reaction to a small increase in the number of hormone molecules binding at the membrane is a cascade of chemical reactions, each increasing the internal response tenfold or more. A speedy, almost digital, response of the cell is the result; because different cells have different thresholds for beginning this chemical cascade, the body as a whole reacts in a graded fashion.

The highly social insects are the beneficiaries of millions of years of positive feedback; natural selection has tended to favor more of whatever works. Success breeds success, as the behaviorally rich have systematically become richer. If we summarize the trends from primitive solitary Hymenoptera to the eusocial bees, we see a pattern of increasing learning, mapping abilities, and social intelligence. Solitary I insects typically lay their eggs on or in a host animal or plant. Because they simply fly about until they encounter cues signaling the presence of a host, they need no learning or communication. Solitary II insects, on the other hand, must build a nest and provision it.

This move from Solitary I wasps to Solitary II requires site learning, and thus the development of at least Tier–3 local mapping from its Tier–2 precursor. But this change now permits the subsequent evolution of cognitive maps (Tier 4). The simple step from mass provisioning to progressive provisioning can require multiple local maps (and a broad-area cognitive map to plot the nest locations). Generational overlap in progressive provisioners may require some modest Social I learning and basic communication, and more of this learning and communication (once present) is bound to be selected for.

The step to Social I organization puts a premium on developing individual recognition, more communication, and flexible nest building. These elements make Social II network organization possible, with its dominance hierarchy and coöperative nest building. Semisocial insects, with their extensive group foraging and context-dependent

Strategies in Hymenoptera

	Provisioning?	Nest?	Site Learning?	Guarding	Social Learning?	Map Learning?	Communication?	Tier/Level
Solitary I (fire & forget)	mass prov	no nest	no site learning	no guarding	no social learning	no map learning	no commun.	Tier 2 Social 0
Solitary II (stock & lock)	mass prov	rote nest	site learned	no guarding	no social learning	map possible	no commun.	Tier 3 (4) Social 0
Solitary III (single parent)	prog. prov	rote nest	site(s) learned	may be guarding	social learning?	map possible	possible commun.	Tier 3 (4) Social 0 (1)
Social I (parallel play)	prog. prov	rote nest	site learned	some guarding	possible hierarchy	map possible	some commun.	Tier 3 (4) Social 0 (1) (2)
Social II (tough love)	prog. prov	flexible nest	site learned	more guarding	dominance hierarchy	map likely	commun. essential	Tier 3 (4) Social 1 (2)
Social III (semi-social)	prog. prov	flexible nest	site learned	full-time guarding	dominance or castes	cognitive map	more complex commun.	Tier 4 Social 1 (2)
Social IV (eusocial)	prog. prov	complex nest	site learned	full-time guarding	castes	cognitive map	elaborate commun.	Tier 5 (6) Social 2

division of labor, depend on even more communication, complex decision making, and cognitive mapping. And all this would seem to open the door to language, vertebrate-level learning, concept formation, elaborate flexibility in decision making and implementation, and the complex architectural accomplishments of this group.

In short, smart has caused selection for smarter, and each expansion of mapping from one tier to the next has opened new intellectual niches. These have almost inevitably depended on increased sociality, at once the cause and result of cognitive flexibility. Speciation has been greatest in the least mobile groups: ants and termites. Among bees and wasps, the most social and neurally gifted species have spread across and between continents. It's a fact of life that intellectual ability broadens an animal's niche (look at humans, rats, and crows), so this pattern of ever fewer, ever smarter species should not be a surprise.

But an equally evident pattern among insects is that intelligence, sociality, and niche breadth have evolved in step with nest complexity. A well-constructed home broadens the niche by virtue of insulating its occupants from the weather and predators. And more social species can (and must) put more effort into constructing long-term homes. To build a nest when there are hundreds or thousands of builders also demands higher levels of cognitive control and decision making by individuals, as well as more flexibility and wholly new strategies for coördination. Again, more both allows and selects for more.

There appear to be two evolutionary positive-feedback loops at work, together and separately amplifying social and general cognitive abilities. Beginning with larger colony size, the social intelligence loop first encounters the necessity for dealing with more social interactions. Selection for increased coördination, sensitivity to current and extrapolated conditions, and flexibility in shifting jobs leads to greater social intelligence—cognitive power focused on social interactions. An increase in social intelligence, in turn, permits larger and more elaborate nests not only to be constructed but also

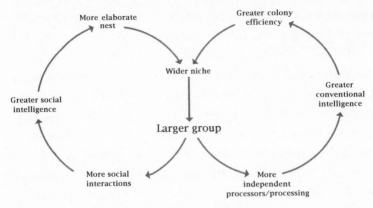

Positive feedback. Selection operates in two parallel cycles to increase group size and cognitive capacity. Social intelligence (left) must generally increase as group size becomes greater, creating more social interactions. Greater social cognition in turn makes possible more elaborate nests, which can widen the species' niche, which completes the circle by allowing group size to increase. In the cycle for conventional cognitive abilities, more individuals means more independent processors and processing, leading potentially to greater colony efficiency, a wider niche, and thus larger colony size.

to be built with greater efficiency. This advance can broaden niche width, including an increase in latitudinal range (a result of better insulation), a wider range of local building sites that can be used, better defense against predators, more efficient rearing of the young, more elaborate and flexible division of labor, and so on. Any of these changes can remove limitations on colony size, leading to larger social groups. And thus this positive feedback allows the evolutionary loop to start again.

The conventional intelligence loop works independently, but in an analogous fashion. More cognitive power, whether devoted to local mapping or planning or to concept use, potentially increases the efficiency of the colony; this improvement, in turn, broadens the niche and leads again to larger group size. We expect to see both these patterns on perhaps a greater scale when we look at the social insects with the largest colonies on the planet (and thus the greatest opportunity for evolutionary reinforcement)—the termites.

TERMITES

High-Carb Diets

There are about 2,200 species of termites (as compared, say, to 3,100 kinds of amphibians or 4,100 mammals); it's one of the smallest orders of insects, and yet one of the most diverse and resourceful. Most termites are tropical; those in the temperate zone are largely confined to North America. Termites make their living from cellulose, the world's most abundant carbohydrate. But though cellulose is nothing but an endless string of glucose molecules, as is starch, it is generally indigestible in the presence of oxygen. The bond joining the sugars of cellulose cannot be attacked by conventional enzymes, which is why it is such a great building material for trees and other plants as well as excellent roughage for most animals.

Termites do not digest cellulose directly, any more than antelope and other grazing mammals do; instead they collect vegetation, chew it up, and leave the chemical breakdown to other organisms. There are two strategies. The most primitive termites swallow the chewed vegetation and pass it to a fermentation chamber where anaerobic bacteria or protozoa break down the cellulose. The termites are nourished by the ever-growing population of microörganisms in their guts that have turned the grass, leaves, twigs, and branches the insects ingest into glucose. Cattle do the same thing: They allow bacteria to ferment the cellulose in an airtight rumen, and then digest the bacteria.

Termite evolution displays several obvious trends, from primitive species very like roaches, which live in small hidden colonies, to the groups a million or two strong, the builders of enormous castles that allow for heat and gas exchange. The less advanced groups, as we just saw, digest some of their pet microörganisms, which do the real work. The culture of cellulose digesters is passed along through a special ceremony. (Termites undergo gradual metamorphosis:

Tiny, transparent, perfectly formed termites hatch from the eggs and totter about begging for food; they grow through several molts in the colony before reaching their mature size.) Young termites feed on a special liquid secretion provided by adults, rich in the group's digestive heritage. When reproductive termites leave, they carry in their stomachs the flora essential for the digestive success of their offspring. Treat a colony of these termites with an antibiotic solution, and they will slowly starve to death.

More advanced species have a different feeding strategy. The energy source is still cellulose, but it is digested outside the termite's body. (Not having to lug about a large chamber of slowly fermenting cellulose solution makes these species more nimble and efficient.) Foragers bring twigs and leaves back to special chambers and chew them, then transplant bits of fungus growing on other pieces of nearby vegetation onto the gnawed edges. Fungi is the only kingdom of organisms able to digest cellulose in air, though they need warmth and humidity to do the job efficiently. This is just what the termites provide. Moreover, these social insects carefully weed the fungal garden and then treat it with antibiotics they secrete to keep bacterial growth to a minimum. When it is time for the fungus to reproduce, pieces are carried into the open to complete the life cycle. Some species are found only in termite mounds, frequently those of a particular species; without their caretakers, they would die. Needless to say, the termites eat the fungi; neither can live without the other. Reproductives carry a chunk of fungi as a dowry when they leave on mating flights.

The evolutionary trend in termites is to forsake excavated nests in soil or wood, like those of most ants, for carton nests either inside excavations or on trees. ("Carton" in termites is defined rather broadly because termites can mix adhesive saliva or feces with pulp or earth, and even sand, to create cells, floors, walls, graceful arches, tiered roofs, chimney stacks, and buttressed towers up to twenty feet high.) Primitive termites do not store food; they live from hand

Termite carton cells. Primitive termites tunnel through wood or soil, but more advanced species build nests. This termite nest includes many separate chambers built with shared carton walls, each with one or more small doorways. The cells are about an inch long.

to mouth, inside a rotting tree, for instance. Advanced termites have special carton chambers for food they hold in reserve; these supplies consist of nonperishable material such as grass clippings, analogous to the hay and straw fed to cattle in the winter, and are kept in a dry carton loft. Primitive species need wet cellulose, such as damp wood; more advanced species can also process dry material.

To expand their niche in this way, dry-diet termites require a source of water. In arid habitats, the insects excavate vertical tunnels down to the water table as much as 150 feet below, which fan out at the base to increase the area of contact and thus maximize the rate of subsurface water accumulation. Finally, less advanced termites live out their lives in tunnels and cells excavated in or near wood; the more complex species, on the other hand, forage from a central nest; so that they can work in safety, they burrow shallowly through the earth or build mud-covered tunnels on the surface of the ground or trees and around the food they wish to harvest. Carton nests on the sides of trees, with gallery tunnels for the foragers radiating out, are a common sight in the tropics.

Termite tree nest. This species of Caribbean termite has built its mud nest in the crotch of a tree (A) from which enclosed runways (B) radiate to feeding sites.

Termites are creatures of the dark. Like the army ants, aside from their reproductives these remarkable organisms are blind. Their homes, some of the most impressive architectural creations on our planet, are erected in darkness by independent creatures a twelfth of an inch long. They can be more than twenty feet high; scaling their size to ours, the wells penetrate the equivalent of more than twenty miles, and the towers loom three miles high and spread five miles across. And termites must be remarkably fit: water gatherers make several round trips to the bottom of the well each day.

Air Conditioning

All termites are fascinating in their own ways, but we will look only at the species that build structures up from the ground. These edifices serve a common need: to carry away the dangerous levels of heat and carbon dioxide that accumulate in a building containing perhaps hundreds of thousands of individuals, as well as fungal gardens that consume the size-scaled equivalent of 50 to 100 billion pounds of compost daily.

A typical advanced nest has a royal chamber, nurseries, gardens, waste dumps, a well, and a ventilation system. The royal chamber is an astonishing piece of work. It contains the queen (or perhaps two), grotesquely swollen with eggs. There is also a king, who, though huge by termite standards, is dwarfed by the queen who may be five or six inches long and an inch and a half across. Unlike the Hymenoptera, the female reproductive termite mates with one or more of the resident males whenever more sperm are needed. The ovoid chamber in which reproductives are sealed has small holes for ventilation and for allowing the workers access. Nonreproductive termites are so small that enemies such as army ants cannot pass in. Workers transport food into the chamber, and carry as many as thirty thousand eggs a day out to the nurseries.

Different species place the royal chamber inside a larger nest, or at the periphery, or at the end of deeper tunnels. A similar range of strategies is employed for laying out the nest. Some termites focus entirely on a large central structure incorporating gardens and nurseries, whereas others disperse them to multiple locations throughout the network of tunnels. The most impressive and beautiful nests are highly centralized apartment buildings constructed with one or more vertical "staircases" to allow rapid movement up and down. The floors may sag gracefully, or, in other species, they can be as flat and unyielding as poured concrete.

The outside of this ovoid bunker is perforated by a series of vents or tubes (or vents converging on circumferential tubes giving rise to more vents, or an arrangement even more elaborate); the structure of these vents and tubes is so unique that they are often used for species identification. As a rule, the vents run down from the inside to the outside, which would keep dripping moisture out and draw cool air up and into the structure. The entire home is suspended from all walls on arching pillars. Ventilation shafts bring cool fresh

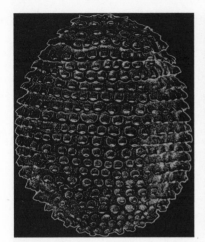 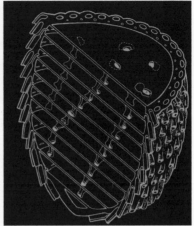

Ovoid termite nest. (A) This small nest is about six inches in diameter and is built out of the insects' own excrement in a large cavity underground. The array of downward perforations allows for heat and gas exchange between the surrounding cavity and the nest. (B) In cross section, the regular multitier structure of the nest is visible, along with the several sets of "stairways" that allow the insects to move quickly through their home.

air in and carry warm stale air out. Tunnels provide access to the surface.

Other nests have a looser apparent design. *Macrotermes bellicosus* (*natalensis* in older literature), which is found from the Ivory Coast on the Atlantic across Africa to Ethiopia on the Indian Ocean, has been studied in some detail. The occupied part of the nest is constructed entirely of carton, and held several inches away from the soil and subsequent superstructure by pillars and arches. Inside the nest, the royal chamber is suspended on arches near the center, surrounded by tunnels to an encircling array of nurseries and gardens. The space between these cavities, and between the chambers and the wall of the nest, are filled with ventilation tubes, some access tunnels, and a large number of insulating chambers visually identical to those we have seen in social wasps and some bees.

Beyond allowing for homeland security, multiple nurseries and gardens, and a well, larger termite nests desperately need ventilation.

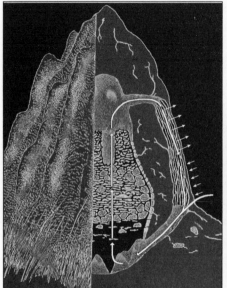

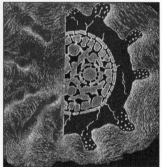

Termite nest. (A) This *Macrotermes bellicosus* nest can grow to fifteen feet above ground level. Most of the structure is given over to gas and heat exchange. The central core of cells contains up to several million termites and many pounds of fungi digesting vegetation. The heat and carbon dioxide generated by the insects and fungi rises through the core and then diffuses out through one of the many buttresses. As warm, CO_2-rich air leaves, it is replaced by cool oxygenated air, especially nearer the ground. (B) From the top, the ring of buttresses is particularly evident. The circular structure near the middle is the royal chamber.

In the dispersed designs, each chamber, especially the gardens, may have its own chimney; some include an earth-carton cylinder protruding a foot or two above the surface. The centralized designs typically have the most elaborate ducting. It is hard to appreciate the organization of the air-flow systems from simple cutaways. A naturalist overcame this conceptual problem by pouring more than eight hundred pounds of fine watery cement down the chimneys of a nest until the entire system was full. When the cement had dried, he chipped away the carton. The resulting cast showed in detail the main chimneys and the radial connections slanting upward from the central cylinders to the peripheral ones. Less obvious was the network of narrow return ducts connecting the outside top to the outside bottom.

Heat generated by the termites and their gardens in the core of the nest flows into the collecting pipes and rises in the chimneys at a rate of about five inches per minute. As this humid CO_2-rich air flows up the chimneys it draws cooler air in through the cellar area under the nest, where it begins to flow up into the various chambers. Depending on the nest design, the hot air may simply be vented by chimneys, or it may circulate down peripheral buttresses. The buttresses are riddled with tiny holes too small even for the termites but large enough for the warm stale air to diffuse out while cooler fresh air percolates in. The buttress strategy retains more of the humidity than the free flow of exhaust air up open smokestacks, but the rate of exchange is slower.

Redrafting the Plans

In fact, *M. bellicosus* can build different sorts of air-exchange systems. One strategy uses no exchange buttresses, but sends the warm air directly to the top to be vented out. Whether this is the result of genetic differences from one part of Africa to another, or a provision in the genetic program for multiple designs depending on the local climate, or whether it might be some group-level consequence of individual response to local contingencies in the nest is simply unknown. But the capacity for major redesign is not restricted to Africa; the Australian species *Nasutitermes triodiae* builds nests twenty feet high in several different styles depending on the ecological zone. Again, the source of this presumably adaptive variability is a mystery.

One hint that the architecture may, at least in part, be the result of independent workers making local cost-benefit decisions comes from tests that impede the normal working of the air conditioning. When one researcher enveloped a nest in a plastic tent, the termites

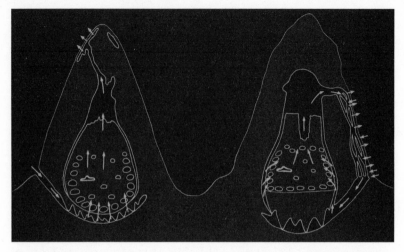

Termite nest variation. The nest of *M. bellicosus* from the Ivory Coast (right) is the one shown in more detail in the previous illustration. In Uganda, the nests depend far less on diffusion: There is a substantial entrance at ground level for cool air and a main chimney that leads to a highly perforated area near the top, where warm CO_2-rich air is released.

responded within two days by building a set of soaring, highly porous cones reaching up from the previous structure and attached to its ventilation columns. The cones were unlike anything ever before seen in the species.

Another way to look at this question is to vary the ecology of a nest while it is being built, or to damage the structure. *Cubitermes fungifaber* termites live in the African rain forest. Their nests can be above the flood plain on a pillar, or not, depending on need; ventilation is managed by wonderful pagoda-like chimneys. The number of pagoda roofs varies, apparently in response to rainfall: more rain, more umbrellas. The progress of building, however, is less easily explained. Sometimes a colony will build a tower equipped with several roofs, and then start another adjacent to it that will not get an umbrella. Other times the first tower will have a roof, then a second tower will be erected with no pagoda, and then a third tower with a roof. Some colonies will build two attached towers and then either double-roof one, or give one a pagoda and erect its twin beside it. This is not carelessness: there are no half roofs.

Hundreds of thousands of blind workers are somehow coöperating in all this; the work goes ahead as rapidly as though there were an architect and foremen, but the plans seem to allow multiple quite different buildings. This looks like an outcome that can be managed only through the kind of dynamic control (or creative conflict) we saw at work in honey bee building, swarming, dancing, and temperature control—the real-time balance that doubtless runs the machine in less easily studied social insects. Each animal, based on its personal threshold for performing a task, the local conditions, the current status of the building site, and the array of innate and learned motor programs at its service, makes a choice. Collectively, these decisions—and more important, the consequences of these decisions—shape the subsequent building. Or is this the whole story?

Added to this mix is variability of unknown origin. Although all honey bee colonies look pretty much alike (with allowances for the cavity the comb is being built in), any two termite mounds built by the same species may be only vaguely similar. Although we can try to explain the difference in terms of local conditions, this approach collapses when we look at the ways the termites repair the mounds. Whatever has led to a particular combination of tower and roof, the response to vandalism is at least as variable. Take a one-tower/one-roof structure and cut straight down from the top center, removing exactly half of the pagoda. Continue the cut down a bit more (about a third of the roof height), and then cut straight out across the cylinder. The result is a column that rises normally, then suddenly half of it and the roof above are missing.

Sometimes the termites close off the open top of the cylinder, cover the exposed side of the half-cylinder above, and then repair the roof, creating a new pagoda symmetrical with the intact upper half of the column. Or perhaps not. They may instead seal off the cross-sectional openings of both cylinder and roof, and start a new roofed tower next to the original one. And then there are the colonies that abandon the carefully mutilated tower and start over some distance away:

If this seeming randomness is the result of multiple local, independent decisions by worker termites oblivious to the large-scale design, then there must be some strong positive-feedback loop that amplifies early choices among equally likely alternatives. This is the termite equivalent of the butterfly in Japan, whose random flight from one bush to another can (according to a popular misinterpretation of chaos theory) start a chain of events leading to a hurricane half a world away. But given that social control in nearly every other context has a self-limiting negative-feedback element built in, this would be quite odd. Perhaps the most dramatic instance of positive feedback is seen in colony defense. The odor released when a honey bee stings induces several nearby bees to sting; the odor of those stings recruits an even larger number.

Assembling Arches

The positive-feedback explanation seems to fail when we look at building on a more local scale. The same variability is seen at the individual level. Whereas our funnel-building wasp ran though a routine exactly the same way again and again until the task was complete, social insects are far less compulsive. The basic building step in many termites involves gluing fecal pellets to make arches; the arches, supporting a network of other arches, provide most of the structural strength needed to support specialized chambers, ventilation shafts, and insulating cavities, and they supply convenient walkways as well. Recycling feces is a superb way to turn a problem

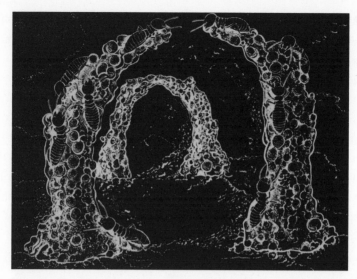

Termite nest arches. The basic architectural element of many termite nests is the arch. These structures are built coöperatively in the dark from mud and fecal pellets. The spacing and curvature of the pillars impose special challenges on the termites as they attempt to complete each arch.

into a solution. But when we look at the pellet sequence, we can already see the sort of individuality that makes it hard to think of social insects entirely as robots. Sixty-four percent of the time the termites bring a new pellet to the building site, fit it in place, remove it, apply glue, then fasten it in. Twenty-four percent of the time they select a spot, apply glue, then fetch a new pellet. Nothing robotic about that variability.

The construction of the arches goes well beyond flexibility and variation. Researchers once thought that individual termites wander around dropping pellets at random, but were more likely to add to a preëxisting pile, thus beginning a column. But we already know that pellet placement is more premeditated: Many termites place glue first, rather than a pellet. And if the random-placement model were correct, then the spacing of columns should be equally random—but it isn't. Columns are neither too close nor too far apart to permit the subsequent construction of arches. So though there may be some social facilitation once a pellet cylinder is under construction, the dumb-chance model for initiation seems inadequate.

Once columns are started, they must be joined. How do these blind animals construct gradually sloping sides at the right height (which has to differ depending on separation) and in the right direction, all in the dark? Researchers have speculated that olfactory cues might be involved, that currents of rising warm air might waft odors from one growing tip to another. But this scheme works only one way. Only a downwind column could arch toward an upwind one; all arches would be built along the axis of air flow, and would have a shape not seen in nests.

air flow ⟶ ▯ ⌐

The behavior of the termites as they undertake the construction of an arch network is to walk to the top of one column, then to a nearby one, perhaps back again or to another pile, and then to place the pellet. This pacing, which is very like what we see in bees when they decide on and lay out nests, may create a third-tier mental picture of the relative location, height, and direction of the columns. Indeed, it's a much more plausible basis for deciding where to start the support cylinders. Only columns within a particular range can be effectively bridged, and the height and curvature depend on that initial separation. The building is more efficient, indeed, only possible, if the workers have some sense of space and necessity. Whether the termite imagination extends beyond the arch at hand is less clear. Could an effective set of ducts be built entirely on the basis of local cues, even with graded responses and variability? Does it require a fourth-tier map? Does it involve social-network maps? These are not questions most scientists thought worth asking until a decade or so ago, and it seems that no group of animals is so loath to share its secrets as termites.

The model of dynamic control, innate and learned motor programs, network and cognitive maps that has emerged from the social insects will guide our thinking about the vertebrates. It's hard to come up with a direct human analogy, our species being instinctively incompetent architects. But the phenomenon might work something like walking. We like to say our children learn to walk at a certain age, but humans no more learn to walk than fish figure out how to swim, or birds reason their way through flight school.

The rhythmic alternation of leg movements is there at birth; a newborn supported under its arms will "walk" on a delivery table. The reflexes that shift weight from one leg to another are there as well. But the muscles are too weak, and the nervous system has yet to calibrate the sense of balance in the inner ear, for the child to discover its ever-changing center of gravity, and to perfect the relationship between muscle-nerve firing and muscle contraction. We walk by taking advantage of a graded response of muscles to stimulation (different fibers have different contraction thresholds), dynamic control (the pairings of muscles that pull in opposite directions), innate motor programs and reflexes, ongoing recalibration of the equipment to the current contingencies, and an all-consuming motivation to master the task. In the end, we consciously decide where to go, use our cognitive maps to select a route, and then for the most part hand over the task to the onboard computer.

This is not to say that the social intelligence that seems to transcend anything in the intellectual armory of solitary insects and spiders is nothing more than a coördinated set of self-calibrating muscle movements guided by a cognitive center. A social invertebrate requires hundreds of such behavioral modules, organized hierarchically, with just the right balance of diversity and similarity if the group is going to work. The question as we turn to vertebrates is whether this may also be the basic strategy some nonsocial but large-brained animal architects rely on.

CHAPTER 6

Bird Nests:
The Modest Beginnings

BIRDS MAY BE the most fascinating group of animals the planet harbors. Their forms, colors, songs, and ways of life are more varied than our imaginations are generally prepared for. They are for the most part diurnal, and thus active when we can see them. And some build nests of astonishing complexity. Birds, not mammals, are the most recently evolved group of vertebrates.

The purpose of classification (phylogeny) is to reconstruct the evolutionary history of organisms based on the characteristics of today's species. All members of a group have a common ancestor: all species in a genus, for instance, derive from an earlier, often extinct species. The animal kingdom is divided into many phyla, including the arthropods of earlier chapters and the chordates, which will now be our focus. Chordates, which are distinguished by having some sort of backbone, are divided into several classes, including bony fish, amphibia, reptiles, birds, and mammals, all of which derive from the humble sea squirt.

The common ancestor of arthropods and chordates may have been something as remote and cognitively unprepossessing as a jellyfish. Thus, we might as well be studying life from another planet when it comes to analyzing insects. To what extent can the patterns of niche, architecture, mapping, and intelligence so evident in insects inform our understanding of vertebrate minds?

The Land Nests of Reptiles

Our fascination with birds and mammals can blind us to the reality that both classes are merely modified reptiles. The reptiles invented the land egg and internal fertilization (as well as upright posture, strong jaws, and a more efficient heart); early birds and mammals merely discovered how to modify scales into insulation—feathers and fur—and thus stay warm. The first birds could not even fly.

Birds are part of the most modern reptile group, which includes crocodiles and dinosaurs; turtles are much more ancient, and snakes and lizards branch off after mammals but before the modern reptiles. We need to look at reptilian nests if we are to understand where birds started their behavioral journey toward elaborate construction projects.

Animals build nests to help protect their reproductive investment. Most vertebrates do not construct anything, since the cost usually far outweighs the benefit. But most reptiles do at least take the trouble to hide their eggs; many either build camouflaged nests or guard the eggs, or both. An ideal parent (from the point of view of the offspring) would also keep the eggs and hatchlings warm and bring food to the newborns. No living reptile does all these things, though some dinosaurs may have. That the whole range of dinosaurs, large to small, was replaced by warm-blooded birds and mammals suggests that controlling body heat has been a key feature in recent evolution.

The strategy of simply hiding eggs is common among contemporary reptiles. Female sea turtles make their way ponderously up a tropical or subtropical beach, excavate a hole, and lay a large clutch of eggs; after covering the chamber with sand, they return to the ocean to feed and breed. The eggs, meanwhile, are warmed by the sand while being protected from predators and direct sunlight.

Some snakes do more. They select a hidden spot, lay their clutch, and then encircle it. Some species warm the eggs as well: the mother shivers her muscles, generating heat in the process. This primitive kind of incubation is a metabolically expensive endeavor, and is reserved for the coldest times; as a result, the embryos have a kind of thermal safety net, but otherwise depend on ambient air temperature for speeding their development along.

Some species of crocodiles excavate a chamber; a few others use a preëxisting one near the shore. They lay their eggs with an eye to warmth, but, as with turtles, fill the chamber in and let the sun-warmed ground do the incubating. Other crocodiles pile up a mound of vegetation and dig a burrow; the fermentation of the plant matter then warms the eggs. But whether it uses a mound or a chamber, the adult crocodile usually remains in the vicinity keeping watch on the clutch for the two or three months until it hatches. Some even help the youngsters out of the nest. But beyond that, the hatchlings are on their own. What we do not see are crocodiles excavating a cavity, laying eggs, and then guarding the chamber.

Our information on dinosaur parenting is, needless to say, indirect. Clutches of fossilized eggs have been found, but that is what we would expect of a reptile anyway. Some species laid their eggs in mounds, others in excavated sandy pits. Certain dinosaurs, such as *Oviraptor,* seem to have sat on their eggs: some parents have been found fossilized in an incubation posture atop nests. Whether this was a matter of guarding the reproductive investment or actually warming the eggs is another question. Active cold-blooded animals with dry skins will inevitably heat up to some extent as a result of muscle exertion. Theoretical calculations suggest that the larger dinosaurs probably did not cool off completely at night. Though *Oviraptor* was only the size of a (very) large dog, it would probably have retained some heat that it could transfer to its eggs.

There is also evidence that some species reared their young in nests. If, as with crocodiles, the hatchlings left the burrow or pit all at once and at the first possible opportunity, any that happened to be fossilized in place before they got out will show no wear and tear; their teeth will be smooth, their bones and joints free of the marks of use. If, on the other hand, the young remained in the nest for a time and were fed by the parents, the telltale signs of age will show up in fossilized nestlings—and this pattern is just what has been found in some kinds of dinosaurs. Although we cannot know about camouflage, the earliest birds may have inherited advanced reptilian behavior for building, guarding, and warming the nest, and then feeding the young.

MEGAPODES

Attempting to use morphological, physiological, developmental, and genetic characters to reconstruct evolution is at once tricky and essential if we are to make sense of how behavior and the minds that generate it have evolved. Biologists arbitrarily divide life into a hierarchy of seven levels running from kingdom to species (eight if you group kingdoms into domains). We've already seen that birds are considered one class of vertebrates. Classes are themselves subdivided into orders, orders into families, families into genera, and genera into species. Birds are divided into roughly thirty orders, and the sequence in which they appeared is approximately known.

The phylogeny of birds presents surprises. Who would have guessed that owls are closely related to hummingbirds, or that albatrosses and boobies are quite recently evolved? Species are not evenly divided among groups; just as insects dominate the arthropods, passerines (perching birds) have been so successful in filling niches and diversifying that they now account for about 60 percent of all bird species. The passerines, which include all the songbirds, also have the most complex nests created by vertebrates. Passerines

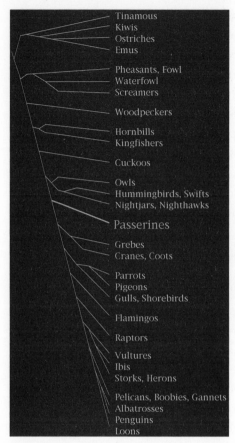

Tinamous
Kiwis
Ostriches
Emus

Pheasants, Fowl
Waterfowl
Screamers

Woodpeckers

Hornbills
Kingfishers

Cuckoos

Owls
Hummingbirds, Swifts
Nightjars, Nighthawks

Passerines

Grebes
Cranes, Coots

Parrots
Pigeons
Gulls, Shorebirds

Flamingos

Raptors

Vultures
Ibis
Storks, Herons

Pelicans, Boobies, Gannets
Albatrosses
Penguins
Loons

Bird phylogeny. Genetic similarities suggest this set of relationships among the birds. According to this taxonomy, the most ancient group of birds with modern descendents is the family that includes most of the flightless birds; many of these species are from Australia and neighboring islands. The next oldest group includes pheasants and other fowl, including the megapodes. The majority of species are in the passerine family.

include many familiar families, including flycatchers, shrikes, vireos, crows and jays, larks, swallows, chickadees and tits, nuthatches, creepers, wrens, warblers, thrushes, thrashers, starlings, New World sparrows, Old World sparrows, cardinals, blackbirds and orioles, finches, pipits, waxwings, and so on.

The earliest birdlike reptiles appeared perhaps 125 million years ago. With the breakup of the earth's single large land mass (Pangaea) about 100 million years ago, emerging groups began to become isolated; the most obvious example is the collection of ancient flightless birds in Australia.

The group that most interests those who trace the evolution of nests and nesting behavior is the gallinaceous fowl: pheasants, chickens, turkeys, grouse, and especially the megapodes (which means *large feet*). Gallinaceous birds can fly short distances for safety, typically abandoning the ground for trees when danger threatens or night approaches. All nest on the ground. Megapode nests are more like the egg-protection projects of reptiles than those of any other group of birds; and although all other fowl incubate their eggs, megapodes obstinately refuse to share their body warmth with their developing embryos. The lengths to which they go to avoid the seemingly obvious solution are sometimes epic, and say something about the mental equipment brought to bear on the problem of nests and nesting from a reptilian heritage.

Megapodes are found in Australia, New Guinea, eastern Indonesia, and nearby islands—all tropical habitats. The dozen or so members of this family run the gamut of parenting styles from turtles to crocodiles, but all use their large feet for digging. The maleo fowl, for instance, live in tropical forests, but travel to warm beaches to breed. Maleo hens coöperate to dig pits about five feet in diameter and two feet deep. They lay their eggs, cover them with warm sand, and leave.

There must be some notion that warmth is critical, because when individual females discover heated soil near volcanoes or hot water streams descending from volcanic peaks, they may opt to dig and lay there rather than on the more usual (but more distant) beach front. This flexibility indicates that the nesting sites cannot be strictly imprinted. The reptilian heritage is evident on the beach: sea turtles and megapode chicks sometimes crawl up and out on the same day, the baby reptiles struggling into the sea while the chicks set off uphill toward the trees. The hens returned to their native forest many weeks before the eggs hatched, so parental care is absent; indeed, megapode hatchlings are the most precocial (fully developed) of all bird chicks, and are said to kick, rather than pip, their way out of the egg.

The remaining megapodes are mound builders. Among th,
the scrub turkey, which fabricates the largest of all bird nests, fro.
thirty-five to forty feet across and fifteen high. It occupies a habitat
between the tropical forests of brush turkeys and the arid open bush
of malee (distinct from maleo) fowl, the two most intensively studied members of the order. These are all "incubator birds": they create a massive compost pile and use the heat of its fermentation to
warm the eggs. But the way the pile must be managed differs dramatically depending on habitat and weather.

The chicken-sized male brush turkey piles up damp forest vegetation into a mound from ten to twelve feet in diameter and from four
to five feet high. He tamps the pile down and adds more until the
size is about right, then monitors the internal temperature each day
until the fermentation process brings it to 95°F. Brush turkeys live in
the forest, so the temperature and other climatic conditions are
moderated by the forest cover, but the male keeps careful track of
his nest. To check the warmth of the pile, he must dig down over his
head. If the incubator is too cool, he adds more compost; if too
warm, he leaves the hole open for ventilation. When the temperature is right, he calls the female to lay an egg in the monitoring pit,
then covers it.

The male continues to check the temperature daily, making the
necessary adjustments. Meanwhile the hen, to whom he is mated for
life, adds another egg every two or three days. The eggs take from
nine to ten weeks to hatch, but since the female can produce twenty
eggs over the course of seven weeks, the male is on duty for about
four months. The female, for her part, leaves once the eggs are all in
the incubator. Despite his effort and apparent concern about his developing offspring, the male ignores the chicks as they struggle to escape the nest, leaving them to seek safety and food on their own.

The four months brush turkey males invest in repetitive nest duty
pales in comparison to the arduous ten or eleven months of ever-
changing manipulations the mallee fowl takes on. When the annual

rains come in his arid habitat he springs into action, digging a pit three feet deep and five or six feet across and then filling it with vegetation from the mallee eucalyptus plant. The basic behavioral element, backward throwing, is the same used in digging, scratching for food, and tossing compost. In this dry habitat all the useful plant matter for fifty yards around may be tossed in. It's essential that the bird finish this task before the rains end, because only damp vegetation will ferment. In exceptionally dry years, the birds do not nest.

Now the problem is to keep the moisture in, particularly during the dry and windy months. The male adds a mound of loose sand perhaps four feet high and twelve feet in diameter as insulation. The coarse vegetation may take as long as four months to start rotting, during which time he digs in and checks the temperature about three times a week. Once the incubator reaches 84°F, the female starts laying eggs. In this sparse habitat, she needs from four to seven days between eggs to find enough food to keep laying. Her twenty to thirty eggs (constituting three times the hen's own weight) are laid over a four- to five-month period. With a seven-week incubation time, the male must attend the nest for ten or eleven months a year, striving for an optimum temperature of 94°F. For several months, the laying is going on at the same time chicks from earlier eggs are hatching and crawling through the sand to get free.

The habitat adds more complications. Unlike the brush turkey's forest, there is no rain to dampen the compost, nor shade to prevent overheating. To make it all perfectly difficult, the nights are often cold. So though the male is digging down through several feet of sand at least daily to monitor the internal temperature, the remedy to overheating or underheating varies. The bird maintains ventilation shafts during the day in the spring as fermentation runs in high gear, but these must be closed if the night is cold. He adds more sand in the summer to protect the eggs from the sun, but scrapes it back at dusk so that the evening warmth can reach the nest. By autumn, the mound is dry and no longer fermenting; now he must re-

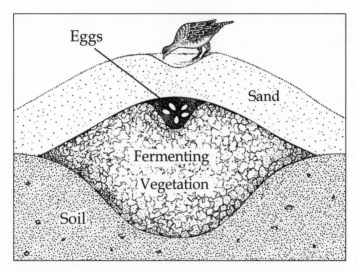

Mallee fowl's nest. Like the brush turkey, the incubator bird excavates a pit, but puts the fermenting vegetation under the eggs, and then adds an insulating cover of sand. The males control the temperature by varying the thickness of the sand layer.

move the sand in the daytime so the nest can catch the sun's rays, and replace it at night to retain heat.

Thanks to the meticulous work of H. J. Frith in New South Wales, we know that temperature regulation is an active process involving real-time measurements and decisions by both members of the pair. Frith placed thermometers throughout the mound, introduced a window into the nesting chamber itself, and added a heater underneath. When he artificially warmed the nest in the spring, the male dug one or more ventilation holes and kept them open for as long as necessary to moderate the temperature. But during the summer, the bird interpreted the heat as arising from insufficient protection from the scorching light of the sun, and so added more and more sand as insulation. This made the problem worse, since the actual source of heat was below. As Karl von Frisch remarked in his commentary on Frith's work, who knows how high the mountain of sand might have grown had not the generator supplying electricity to the heater failed? Finally, artificial heat in the fall induced the male to leave the sand in place instead of shifting it twice a day.

It seems clear that the mallee fowl is actively monitoring temperatures and making sensible behavioral choices. Birds that construct incubators must know innately how and when to commence construction and what materials to employ, as well as what temperature is best; doubtless, too, the formulas for adding material, removing it, and digging ventilation shafts are also part of their inheritance. All this is specific to the species. But frequent decisions must be made about whether to take action, and if so, which response is best. That mallee fowl recognize that the choice depends on the daily cycle of ambient temperature and sunlight, as well as the ever-changing rate of fermentation, is remarkable, and as noteworthy as the fact that the males stick with the problem 24/7 nearly the whole year.

This network of interacting parameters and behavioral options looks like Tier–5 network mapping. Experiments to determine the role (if any) of experience in this species could further unravel the mysteries of this behavioral tour de force.

SCRAPES AND CAVITIES

As we have seen, some megapodes breed like sea turtles, but others employ a wildly exaggerated version of the crocodilian compost-pile tactic. Two other relatively unmodified examples of reptile nesting are found in birds as well, the difference being that these warm-blooded birds brood their eggs, transferring heat and thus helping make the development faster and less susceptible to the problems created by temperature changes. But these reptilian nests are so simple that they may not even deserve the name. One kind is an almost imperceptible natural declivity on the ground (or one created by a few backward scratches); the other is an unadorned natural cavity.

It may seem odd that a book about animal architecture should discuss such nest-challenged species, but they are important to the issues of nest evolution and bird cognition. And since this minimalist strategy serves many familiar avian species well, we might wonder why other species go to the trouble of fabricating their elaborate nests. The birds that make do with scrapes include most pheasants (which share their order with the megapodes), terns, plovers and killdeer, nighthawks (the most nestless of all ground-brooding birds, with vultures a close second), peregrine falcons (an inch-deep scrape), and even short-eared owls (a slight depression on the ground). The evolutionarily ancient ostrich just tramples down the grass where he chooses to nest, scrapes a hollow about a foot deep and three feet across, and collects from five to twenty-five eggs from one or more females. The far more recently evolved guillemots put their eggs into a bare crevice on or at the base of a cliff, usually within a few feet of other guillemot "nests." No depression is needed: the eggs are pear shaped, and so roll in circles rather than off the ledge. The New World jaçana simply deposits its eggs on a lily pad; the eggs float, and after a mishap are recovered with a stereotyped egg-rolling response.

Sometimes the scrape-nesting birds will add a piece or two of grass, perhaps from a halfhearted desire for camouflage; now and then a plover will encircle its nest with an imaginative ring of stones, the logic of which is obscure at best. A few pairs in one species of tern will drag up seaweed at some cost of time and effort, all for no obvious reason. As always, variation from one animal to another in nest design must make us wonder whether we are observing genetic "noise," a one-time event contingent on timing and circumstance, or actual personal preference (reflecting perhaps aesthetics, logic, whim, or indoctrination). In any event, scrape nesters have to choose where to nest, and then decide whether a scrape is needed to enhance a local depression. Even though these are some of the world's most experimentally convenient species, the bases of these decisions are unknown.

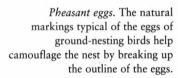

Pheasant eggs. The natural markings typical of the eggs of ground-nesting birds help camouflage the nest by breaking up the outline of the eggs.

Once the nest is established, it is difficult to spot the brooding bird or its eggs even in plain sight. Unlike the eggs of reptiles, bird eggs can be manufactured in a nearly infinite variety of colors and markings; for species nesting in the open, the eggs are invariably camouflaged. The bird sitting on the clutch is as a rule equally invisible, at least while motionless. Perhaps the eggs are difficult for the parent to see as well; some individuals in these species will continue to incubate even if the eggs are removed from the arbitrary spot the parents take to be the nest.

The most brazen of these nesters are the terns, many of which lay their eggs on a scrape in the sand. The fairy tern places a single egg on a tiny depression in a thin branch overhead and incubates in this tenuous position. Field guides describe most tern nests as a "mere scrape" or a "slight depression scratched in sand." As an example of pure chutzpah, however, the potoo (a kind of nighthawk) finds a broken tree trunk, lays an egg atop it, and then incubates in a vertical, head-to-the-sky posture that (given the bird's coloring) makes it appear to be just an extension of the trunk.

Although scrape nests seem cognitively undemanding, keeping the eggs and chicks safe can be another matter. If you nest on the ground, mere inconspicuousness will not be enough when a hunting fox or grazing deer approaches. Based on past experience with the kind of animal, and even sometimes the particular individual, the bird on a nest must decide whether and when to abandon ship and depart. Lots of ground nesters do this, but killdeer and plovers make a point of then trying to distract the approaching animal. If it is a clumsy grazer rather than a predator, the bird basically leaps out in front of it and screams; this is sufficient to cause even an elephant to choose another route. But for the hunter, there are other solutions.

The most well known is the broken-wing display (also performed by jaçanas to lure water snakes away, and by certain other ground nesters). The nesting bird drags a wing, hops away clumsily (often tripping in the process), and in the climax of the performance, de-livered only to the most difficult audiences, even attempts unsuc-cessfully to fly. But plovers and killdeer have other acts in their repertoire as well. All begin with a stealthy redeployment from the nest to a stage chosen on the basis of local topography, taking into account the distance, direction of movement, and nature of the predator. Like all birds with territories, plovers and killdeer must rely on a cognitive (Tier–4) map. Depending on the degree of threat, the bird may choose simply to feign nesting on a high point of ground. Another popular strategy involves rushing through the veg-etation uttering high-pitched squeaks that sound like the cries of panicked (and highly edible) rodents.

The ability to choose wisely from an innate menu of distraction displays suggests an ability on the part of at least some ground-nesting birds to judge intention by behavior, to integrate experience with innate prejudices to create expectations. Here is another exam-ple of multiparameter social intelligence being brought to bear on the challenges of nesting. If these were humans, we would not hesi-tate to apply such terms as "thinking" and "planning," however foolish we think their choice of nesting site. (At least they mostly avoid flood plains, and they evacuate as hurricanes approach.) But their nesting strategy must work pretty well; ground nesters continue to thrive except where our agricultural practices, dune buggies, and dogs have changed the habitat, and thus the basic biological rules.

Another nesting tactic that derives from birds' reptilian ancestry is the drive to find a natural cavity for laying eggs. The problem is to find a dry, camouflaged compartment; after that, the birds using this strategy just deposit the eggs and incubate them. Cavity nesting is generally more successful than ground nesting (66 percent versus 49 percent of eggs hatch), probably because cavity nesters are less often disturbed by predators. As we shall see, the insulation in more elaborate nests raises the hatch rate still more. In some species, these birds are discovering the value of nesting strategies used by social insects for tens of millions of years.

Any number of species lay their eggs in empty unpadded cavities. One of the most common is the barn owl, which often lays on a ledge in the corner of a building. Shelducks search out old rabbit burrows, and six other species of Northern Hemisphere ducks do something similar. The sparrow hawk and screech owl use natural cavities in trees, or the eaves of barns. Barred owls are a bit fussier, disdaining human habitations. No elaborate multifactorial studies have been performed (as they have on bees) to titrate out the relative importance of the many parameters these birds must consider, not to mention the processing that allows this information to be used. But there is a passiveness here, a failure to modify in any substantial way what chance provides, which suggests little mental horsepower is at work. Cavity-choosing birds might need only local-area mapping to select the best spot; a bird that renovates the chamber needs that and more.

A glance at the outline of bird evolution suggests that simple nests are very common among the orders that branched off before passerines. There are examples of extreme elaboration, as we shall see, especially in waterfowl and hummingbirds, but it is with the passerines that nest designs and techniques multiply, and almost no species seems content with something simple. Bird nests in general have become more complex simply because selection has rewarded birds with better nests—better at protecting the young from the elements, predators, and accidents. Often this complexity, and the cognitive equipment that evolved to implement it, has widened or opened new niches for the species. But this is not a process of inevitable progress; when the contingencies of niche and habitat have permitted, returning to the simple answer has remained an attractive option. Among later groups, regression has frequently occurred, especially among shorebirds, some raptors, and penguins. Evidence from a few species of birds accustomed to building complex nests suggests something about how this simplification might have begun: When provided with potential shortcuts, some birds have the wit to recognize and opt for the low-effort option.

So what are the costs and benefits of more elaborate nests? Perhaps, as real estate agents love to say, it's location. To get off the beach or forest floor, a bird must generally build at the least a substitute for the ground. To overcome the perennial shortage of cavities, new ones must be excavated. Insulation must also be considered: The breast of the incubating bird is almost always much warmer than the ground or cavity. As a result, the tops of eggs are warmer than the bottoms, which cannot be good for the developing embryo. If the bird can create a layer of air-trapping insulation between the clutch and the substrate, the eggs will be more evenly heated. Experimentally removing insulation from nests lengthens incubation time, reduces hatching rate, and requires the brooding bird to increase its metabolic activity to compensate.

But full-blown nests are very, very expensive. One way to measure this is simply to calculate how far a bird must travel to gather the materials. While working in Trinidad, William Beebe dissected a mockingbird's nest into its 542 elements. His intimate knowledge of the area allowed him to determine that the birds had needed to fly a minimum of 185 miles to collect the pieces. Another researcher showed that cliff swallows commute 150 miles to collect the 1,400 mud pellets their nests require. Village weaverbirds put in 230 miles to create their remarkable hanging nests. These are extraordinary journeys on a human scale, even allowing that flying is only about 20 percent as energetically expensive as walking. We can take it for granted that fourth-tier cognitive mapping is at work to make efficient nest-material collection journeys possible: the bird first consults its internal map of the home range to pinpoint the source of whatever building components are needed at the moment, and then it sets off along the most direct route.

As more effort is invested in building, clutch size drops; these two activities exact a tradeoff in time and energy investment. All things being equal, then, birds should minimize building costs. In the search for the best compromise between costs and benefits, natural selection has tried many half measures in the form of low-energy changes to minimal nests that will reap some of the benefits of elaborate, high-investment nurseries. The first step was simply to renovate what chance supplied. Does this small extra investment help enough to explain why most species have opted for further domestic upgrades?

RENOVATIONS

One drastic but effective solution to the insulation problem is widespread in the ancient waterfowl group: One or both parents pluck down from the breast and use it to line the nest. Wood ducks and hooded mergansers, for instance, occupy natural cavities (preferably with wood chips or shavings already in place) and add a thick layer of down. Not only does this insulate the clutch from contact with the bottom of the cavity, the behavior *uninsulates* the breast and exposes a "brood patch" of highly vascularized skin that effectively transfers heat to the eggs.

House wren's nest viewed from above. House wrens renovate existing cavities with a series of layers of ever-finer material. Typically, they begin with twigs, followed by grasses, and end with feathers or plant down. Different birds invest to different degrees in these several layers, and show varying tendencies to use substitute materials.

A denuded breast must be inconvenient outside the nest, especially while swimming, but at least these birds don't have to fly hundreds of miles to gather their insulation. Other cavity nesters that adopt renovation half-measures typically bring in material that will pack loosely; though nothing is as good as down, a sufficient accumulation of nearby mosses, dried grass, wood chips, tiny twigs, and even leaves can make a real difference. A fair number of cavity nesters invest in such furnishings. House wrens find hollow cavities in trees and bring in sticks, tamping them down until the layer is thick, and the friction between the sticks' rough surfaces confers stability. Grasses typically come next, with any loose feathers the bird can find added last. Tree swallows stuff dry grasses into a chamber, shaping them into a kind of cup almost incidentally as the birds move around, and then collect feathers for an inner layer.

Nuthatches bring shreds of bark, then perhaps twigs, then grasses, and so on, the material becoming ever finer; depending on the species and the habitat, these finishing components can include moss, rootlets, plant down, feathers, or fur. The trend in nearly every species is toward a sensible order of collection, from rough to fine; the same sequence is used in many of the classic cup nests we will look at in the next chapter.

The prize for modifying a preëxisting cavity goes to hornbills, though the object is safety rather than insulation. Essentially all forty-five species of hornbills occupy natural cavities in trees. The male brings mud to the entrance, which the female uses to narrow the doorway until she can just slip in. Then, using any mud that has fallen into the cavity as well as her own excrement, she continues the work from inside until there is barely enough room for her bill. Older accounts presumed that the male was responsible for creating this convent-like abode, but it is entirely the female's doing, and she is well able to break out at any time.

This modified cavity is a safety device. Two or three dozen times a day the male passes fruit, insects, and small vertebrates through the tiny hole. The female lays her few eggs, and takes this opportunity to molt. The chicks hatch out after about three weeks. Now, with all the extra mouths to feed, the male has to collect about twice as much food. Soon this becomes an overwhelming task, and the female breaks out to help. The chicks immediately rebuild the defensive perimeter.

But the hornbills are an exception; renovation of existing cavities is otherwise a matter of insulation—or at least, that's what it looks like. The home of the great crested flycatcher, however, pushes the limits of this explanation. The two birds gather a motley collection of twigs, leaves, feathers, fur, pieces of bark, lengths of vine and rope, and shed snakeskin. The frequent presence of snakeskin has always puzzled ornithologists; some think that the wrapping of snakeskin around the nests of many blue grosbeaks and tufted titmice, for instance, discourages predators, but the skins of the fly-catcher nest are interior, and thus not visible. And any of these species will collect strips of cellophane tape instead of the real thing. The flycatchers' desire to gather these components reflects an earlier lifestyle when their ancestors built true nests, as most flycatchers still do. Snakeskin—or tape—is a flexible and strong addition to the substructure of a nest that requires external support.

Insulation was not invented only by cavity nesters; it is also seen in some ground nests, where it may have marked a critical break-through into the platforms of more advanced nests. Simple, minimal lining of the cavity is characteristic of turkeys, which are members of the pheasant group; they line their scrape or depression with leaves. Grouse do something similar, and they will add dried grass as well. Many geese carry sticks and other vegetation to their scrape and drop it in. All these species mold the lining from inside the nest by performing the scrape behavior while sitting down.

This carry-and-drop strategy with sticks seems to have opened a behavioral door. It evolved into something more specialized and in-teresting in ground- and ledge-nesting species, such as doves. Pi-geons, for instance, bring small sticks, stand next to the nest site, and just toss the twigs in. The result, given the inaccuracy of the be-havior, is a stick that is loosely stuck into the previous pile of sticks. Surprisingly enough, considering the casualness of the delivery sys-tem, the added element usually winds up oriented circumferentially, which is the best overall direction for nest strength in this species. The roughness of the sticks keeps them from moving freely relative to one another, giving the flat pile an unlooked for cohesion. Giving doves a cache of smoothly turned dowels to work with leads to a hopelessly unstable mess. They do have the wit to select real twigs more often than dowels when offered a choice.

Probably the most elaborate of renovations in ground nests is seen in albatrosses. Pairs of wandering albatrosses use their beaks to scrape out a trench about three feet in diameter. Excavated dirt, moss,

Dove's nest. Doves depend on friction between sticks to hold their platform together. This nest is built on an artificial ledge—the window sill of an academic building in Princeton. This site preference is a preadaptation to life in cities.

and grass is piled in a heap in the center, where it is trampled down. When this hillock is from one to three feet high, the albatrosses make a small hole at the top into which the female lays a single egg. The breeding season in their habitat is so short that without the mound the nest might be buried under snow before the chick is fledged.

Unfortunately, we have no clear idea of what renovators may know or understand. Insulation certainly pays off in reproductive fitness, but it entails its own costs. How much do the species that invest so tentatively in this commodity actually grasp? Do their modest efforts reflect some cost/benefit calculations? And what about their choices of construction parts? The birds seem to be willing to make substitutions: string rather than vines, fine dry grass or plastic Easter moss rather than coarse grass, which shows an adaptive flexibility, perhaps an understanding of the larger goal rather than a slavish need to carry out innate chores in the building process. Or, a skeptic could argue, the innate programs are so vague (so free of selection pressure) that "plasticity" is inevitable.

When the birds grasp the essential features of nesting material, they are doubtless at an advantage: a builder driven mindlessly to gather lichen or moss will be in deep trouble if none is to be found; there is distinct survival value in being able to recognize roots or plant fibers as viable alternatives. Lovebirds, for example, normally collect coarse grass as nest material, but in the lab they content themselves with tearing sheets of white paper into grass-like strips. Could it be that some nest-building birds have a concept of the sort of material appropriate for each step in the process?

This idea would have been laughed to scorn thirty years ago, and yet the first evidence that pigeons can form abstract concepts was discovered during World War II. As part of the war effort, B. F. Skinner set out to discover whether pigeons could be taught to peck at images of ships projected on a ground-glass screen in the nose of a bomb—what would have been the first "smart" weapon. The lab-reared birds, with no experience whatsoever of ships, caught on quickly; once trained, they would peck accurately at photos and movies of ships regardless of angle or distance; they ignored objects that were not ships. The military refused to believe their eyes, and the pigeons were spared a kamikaze fate.

At the time, Skinner imagined some sort of elaborate conditioning was going on. Two decades later Richard Herrnstein, Skinner's student, reopened the issue. He used an approach that resembles the way we ourselves learn about the world. He offered the birds a hundred slides contributed by his graduate students, half of which included a tree. They were in random order, and the hungry pigeons were fed only when they pecked at an image with a tree in it. To his amazement, the birds picked up the discrimination faster than they learn to choose other criteria—color, for instance. When he tested them with a hundred new slides and no food, the pigeons were about as accurate as young children.

Critics countered with the hypothesis that perhaps the birds had an *innate* picture of trees. So Herrnstein tried other potential concepts:

Concepts. Pigeons tested for concept learning were given a series of unrelated slides and rewarded only for pecking those sharing a common feature (+). Later tests to see whether the birds had acquired the abstract concept used new slides. The birds learned the concept shown here (tree) with surprising ease.

female undergraduates, Volkswagen cars, and fish, to name three that are unlikely to be part of the bird's genetic heritage. The birds mastered a Tier–6 concept of each quite quickly, though they were never much good with the concept of household appliances.

What's going on here? First, the pigeons are motivated to learn. In their tests, they were physically hungry; but for humans, the drive is evidently to acquire knowledge and to communicate. From a limited sample, the animals seem to have inferred a probabilistic set of criteria for "tree": usually green, usually with a spreading shape, usually with leaves, usually with branches, and so on. None is a necessary and sufficient feature; each has its own predictive value. This is exactly how humans come to divide the world conceptually into animate and inanimate, chairs versus tables, dogs versus cats, and so on. Sometimes our generalizations and inferences are wrong, but we learn quickly from errors. The result is an ordered world upon which our cognitive processes can operate with great efficiency.

Do birds, then, form concepts about nest material? Are they born with some innate way of recognizing when an object will serve, and the drive to generalize and add that example to their concept for, say, middle-layer, moderately insulating components? Or, to put it another way, if pigeons can create categories about entities as irrelevant as fish, is there any reason they should *not* apply this intellectual skill to an important task? No one has thought of asking them. Another set of challenges for understanding nesting (and the role of network mapping, goal-oriented planning, concepts, and some degree of abstract reasoning) center on what birds know about the larger issues of design. Again, the essential tests—including manipulating the amount of insulation, warming or cooling the cavity or ground site, or providing controlled choices of lining material (to mention the most obvious)—have yet to be tried.

EXCAVATIONS AND PLATFORMS

The next logical step in the evolution of nests is for cavity and ground nesters to adopt a more proactive approach and produce their own cavities or build stable ground-like bases for nests. Let's look at platforms first, and then at burrows.

The simplest technique for generating a better base is to pile material onto whatever inadequate foundation there already is. Grebes, for instance, create floating nests of sodden and decaying water weeds simply by stacking more and more material in a suitable spot. Many aquatic plants have air spaces that enable them to float, and doubtless this helps in nest construction. Starting the pile among some reeds so that it won't be carried off by the current is essential, and these birds are smart enough to use a shallow rise as an island base, or even to loop a stem around a reed to lash the nest in place.

The grebe creates a platform where there was none before, and thus opens up a new nest site. Here is a new niche; a modest architectural breakthrough has made possible the evolution of multiple species of grebes. Bitterns and the magpie geese of Australia do something similar. They trample the vegetation at their marshy nest site, creating a platform. The reeds are bent down systematically from the outside in, so that they overlap and create a strong and stable deck. The water can be any depth so long as reeds project out of it prior to construction. For the geese, this stage serves as a display platform, but then cut reeds are added to raise the level for nesting. Bitterns also create a substantial pile of reeds after the bending-over step, raising the nest well above the water. Both species tend to add a layer of seed heads to the top, presumably for insulation. Again, they have created a platform for a nest where there was none.

The horned coot breeds in Andean lakes where there are few reeds. To take advantage of the safety of depth, the coots must heap up a substantial platform of stones, some of which weigh as much as a pound. Upon this mound, which may be three feet tall and twelve feet in diameter, the coot adds a pile of aquatic vegetation and fashions a dip in the top for eggs. The top of the nest sits about two feet above the water. Some preëxisting behavioral elements probably helped in making trample-based aquatic nests possible, but stone manipulation looks entirely novel. Could an ancestral coot have discovered the behavior by chance (trial and error), or was it through insight? Is the architectural custom innate or learned?

In a very different habitat, many penguins also construct stone platforms for nesting. The birds heap up small stones, tamping them down to produce a structure raised well above the ground. But as with grebes and bitterns, the goal is simply to keep the eggs dry, and, for penguins, safe from torrential runoff during summer storms. The stones are so valuable that theft is endemic, and it involves a degree of dissimulation all too familiar to cognitively overachieving humans. As Louis Bernacchi tells it:

> The thief slowly approaches the one he wishes to rob with a most creditable air of nonchalance ... and if ... the other looks at him suspiciously, he will immediately gaze round almost childlike and bland, and appear to be admiring the scenery.... But no sooner does the other look in a different direction, than he will dart down on one of the pebbles of its nest and scamper away with it in his beak.

When the New York Zoological Society attempted to breed penguins at its Bronx Zoo, it discovered that the birds were so fussy about the stones that thousands of pounds had to be flown in from Antarctica.

These platforms are an important step in the evolution of the classic bird's nest. Once the animals discovered how to create the platform in a tree, an entirely new and much safer habitat dimension opened up. Some platforms are created merely by piling up sticks with an apparent carelessness that makes the sturdiness of the eventual creation all the more surprising. Ospreys and eagles drop branches and sticks from three to six feet long onto likely cliff niches or the tops of trees, counting on friction to hold them together; they add to them each year. One golden eagle's nest in Canada grew eventually to a depth of twenty-one feet, and it contained about four tons of sticks. Another was used for more than thirty-five years. The platforms are sometimes elaborated with brushwood, plants, and moss, which the next year's round of reinforcement (if that is what it is) covers over.

European storks build massive but simple platforms. These nests are up to six feet across, and each year sees a new layer of sticks and then earth. With each passing season the structure grows taller and taller, and those built in trees eventually come crashing down. The nests that have been in use the longest are typically found on houses, which are better able to take the strain than most trees.

The trend in cavity nests has also been toward creating new opportunities. The most aggressive cavity makers are the woodpeckers, a group only slightly less ancient than the pheasants. Though many species are perfectly happy to use an existing cavity, they are born prepared with the morphology and drive to excavate a new one if need be. Other woodpeckers, however, start from scratch. In North America, pileated woodpeckers, yellow-bellied sapsuckers, hairy woodpeckers, and downy woodpeckers take this rather Calvinistic attitude and hollow out their own, reinforcing their austerity by eschewing lining for the nest. Abandoned woodpecker holes provide homes for a variety of cavity-nesting birds, such as starlings, that lack the wherewithal to carve out their own. Thus, the woodpeckers' refusal to enjoy the luxury of a prefab model opens up nesting sites for many other species.

To excavate a new cavity, a woodpecker pair first needs to find a hollow or a diseased tree; sound wood is just too hard to hammer into. A few vigorous taps on the bark tell the birds whether a particular tree is a likely candidate. The woodpecker's beak has a slightly flattened, chisel-like shape that it uses to cut out chips through a series of regular hammerings, the first set directed down. The bird starts from one side—the right, say—moving its head to the left and rotating it clockwise around an imaginary pivot point deep inside the tree. Then the woodpecker does the same thing from below with its beak angled up. Out comes the chip.

The hardest work is often creating the opening, which may pass through sound wood; chipping away at the softer interior goes faster. Different species share the work according to their own plans. In some, the male will put in hours and hours of work, but in others the two birds alternate every thirty or sixty minutes. Some species tunnel up, in an apparent effort to keep the nest dry, and then dig down, expanding their work into a sphere. Others go straight in, build a right-angle turn down, and then fashion the chamber. Some cavities are cylindrical, others are conical, still others oval. There must be at least Tier–3 local-area mapping at work to guide the shaping of the cavity, and the bird seeks the best compromise between its probably largely innate notion of the ideal cavity and the realities of what it has to work with.

Not many other species possess the equipment to excavate in wood. The one exception are the trogons, a small group of colorful tropical birds that eat fruit or take insects on the wing in spectacular hovering flights not unlike those of their close relatives, hummingbirds and swifts. Some trogons excavate cavities in trees, while others create similar chambers in the nests of social insects. The two members of the pair take turns biting at the surface; their less specialized beaks make chiseling impossible, and so the wood has to be soft. Some species create an upward-slanting tunnel perhaps three inches in diameter and seven inches long, and then work down to produce an egg cup. Others proceed in a downward slant, ending after a few inches and then creating the hemispherical nest base. These nests look more like balconies than cavities. Wood soft enough for trogons but hard enough to retain its shape for the breeding season must be rare indeed.

The biting action of the trogons is similar in form to the behavior of some birds that excavate cavities in the soil, particularly along stream banks or on cliffs. Again, the goal is to create a nest site where formerly there was none. Kingfishers constitute an entire order related to woodpeckers and hornbills. There are fewer than a hundred species, mostly found in the tropics, but most temperate areas play host to at least one kind. Kingfishers, which live near ponds and streams, dig a gently ascending passage from two to ten feet long (depending on the species) into a river bank or other convenient slope, adding a nest chamber at the end. Their digging sometimes takes the form of biting, but they also make sharp jabs not unlike the chiseling of woodpeckers, and kick out the dirt chips with their feet. Like essentially all the cavity-creating birds we have mentioned, they do not purposely gather lining material, but the slow accumulation of soft fish bones in the nest provides an ever-growing layer of aromatic insulation. Among the tree excavators, the birds usually omit to remove the last round of decaying wood chips, producing (perhaps accidentally) a soft, low-density pile onto which the eggs are laid.

Motmots, tropical relatives of the trogons with the odd habit of trimming their two long tail feathers down to tufts on the ends, also excavate. The birds begin their nest excavation long in advance of need, then ignore the burrow until the breeding season. All other nest building is undertaken at the last minute, but the soil is softer earlier in the season when the motmot pair elects to build. Their tunnels are impressive affairs, growing to about seven feet in six to eight weeks.

It is from the sequence of steps described in this chapter that complex nest building seems to have evolved. Inherited reptile strategies of compost, cavities, and scrapes that birds sometimes guard and even incubate were superceded by active incubation in preëxisting chambers and surface dips. Modification of the cavities and hollows followed, generally to supply insulation. Finally, some species began actively creating chambers and platforms. Platforms in particular seem to have liberated birds, allowing them to nest in trees. Evident even in some of the most primitive species are cognitive (home-range) maps, concept formation, and network intelligence, all of which offer an array of intellectual tools for construction behavior to build upon.

It's not surprising that we know so little about the ontogeny and variability of burrows and cavities: what kind of nest building could

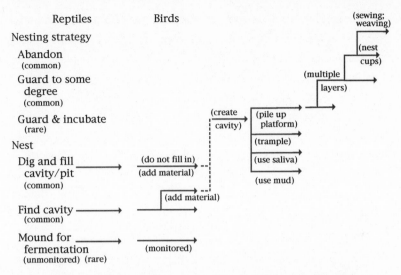

Nest evolution. A very few reptiles guard and incubate their eggs. Among those that do there are three nest styles, all of which are seen in primitive birds. The more advanced bird nests appear to be elaborations on digging and using preëxisting cavities. Most of the strategies for creating complex nests derive from building a platform and then creating a structure firmly embedded in or on this base.

be harder to study? The prospects for understanding the minds behind the nests are much brighter for more elaborate structures built in the open—the next group we will consider. We will focus on how birds have come to incorporate crude platforms and cavities into more elaborate nests, or, more often, how they seem to have taken the concept of platform building and applied it to other "media" to produce structures that continue to strain human imagination.

CHAPTER 7

Bird Nests: Molding, Felting, and Weaving

THE NEST BUILDING PATTERNS we have seen so far extend and elaborate basic reptilian strategies. Birds, being warm blooded, have generally added insulation and active incubation to the other major nest site and nest design criterion: safety through camouflage or inaccessibility. Although the composting strategy has been an evolutionary dead end (that is, it has not led to much speciation), other behavioral and architectural enhancements do represent evolutionary breakthroughs. The invention of wholesale excavation, and of lining the nests with insulating material, are examples. Discovering how to make rude stick platforms, or how to pile up or trample down material to create a nest base, allowed birds to modify and improve opportunistically chosen bits of ground. The possibilities of freestanding aquatic and aerial nests, and the thousands of new sites they made possible, could then be realized.

In particular, though, the evolution of arboreal cup nesting has opened countless new niches, and has led through selection to the concomitant invention of new behaviors and uses of materials.

Positive-feedback loop. Just as with the social insects, cognition in birds is subject to a positive-feedback loop involving niche breadth. Greater cognitive potential permits more elaborate nests, which can enable the species to enjoy a broader niche, which selects for more cognitive potential to exploit the opportunities. To the extent that some of the extra cognitive capacity can be recruited to build better nests, the cycle continues.

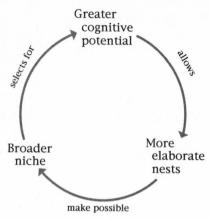

The result is a cycle in which increased cognitive potential not only creates more elaborate nests but is itself pressured to expand. But despite the technical diversity of what is built, the goal of all nest building is the same: to provide a dry, warm (or warmable) place safe from predators. Through millennia, birds have reinvented many of the design features of social insect nests to achieve similar ends.

External Burrows

As with the social insects, the limited number of intact previously owned nests, of suitable banks for digging, or of rotted trees for hollowing out puts a premium on discovering ways to build cavities on preëxisting structures. Cliffs provide a temptingly remote refuge from predators, but their stony banks are impossible to mine with beak and claws. To overcome this difficulty, birds have reinvented and improved upon the mud wasp strategy: they collect mud, often mixing it with grass to make a sort of adobe. The blend can be a fairly strong substance to work with, and the resulting nests may last for years if they are protected from rain.

Despite the advantages of mud, only about 5 percent of birds use it in their nests. In part, this is because reliable sources of suitable mud during the nest-building period are not always available. A more serious difficulty may lie in gathering the mud. This distinctive behavior seems at first glance to be unlike anything in the bird's natural repertoire, except, perhaps, carrying away the fecal pellets of the chicks, or some of the behavior needed for excavating soil. The mud application techniques also look novel. Behavioral evolution works best when it can recycle existing bits of behavior, reordering extant motor program elements and revising the sign stimuli involved in triggering them. Mud use cannot have developed easily. And, as we shall see, a major cognitive barrier must be overcome.

Often where mud is used, it may be only a minor component, as in robins' nests. Other birds, such as flamingos, pile up mud and then sculpt a cup. Depending on the flamingo species, the resulting structure may be a foot or two high and about the same distance across at the bottom (and half that width at the top), and weigh from 45 to 115 pounds. But for these water birds mud is readily available.

A serious problem with mud is that it takes time to dry; it's rare for unsupported lateral or cantilevered construction to proceed more than about half an inch a day; adding more material would pull on and deform the previous layer if it's still wet. On the other hand, waiting too long leaves the bird with the problem of trying to add wet mud to dry; such a joint is inherently weak, and no reinforcing grass bridges the gap. In short, this is a tricky and demanding undertaking, depending on the nature and quality of the raw materials, the site, and the weather. If you are in the market for a humbling experience, try it. You'll need at least a Tier–3 local-area picture of your desired product; you will probably also want a Tier–5 cognitive network for deciding among and orchestrating options, and lots of practice.

Building an external burrow on rock (or, increasingly in many parts of the world, concrete) poses another problem—how to get the construction material to adhere to the base in the first place. Here the bird must judge the mud's affinity for the substrate in advance of the pressure that the fully built nest will put on the junction. As most of us know from experience, the force a glue joint can withstand is related to the area of contact, the strength of the adhesive, the porosity (usually) of the substrate, and the weight to be supported. Birds that build with mud seem to have worked this out, and they increase the contact area either by widening the nest walls or by choosing a spot where the cliff (or bridge, or house) has a right-angle projection to which the roof of the cavity can be attached. Indeed, some species do both. Almost certainly this ability improves with practice. A capacity to understand the goal and how to recognize progress in achieving it seems essential, but there is no good evidence on this point.

Swallows and Martins

The most familiar mud-cavity builders in North America and Europe are the martins and swallows, both members of the same family of birds, and the evolution of their nest building behavior can be traced through molecular sequence analysis of the different species. The oldest branches of the family simply excavate burrows in muddy or sandy banks; a typical example is the bank swallow, which kicks soil out of the holes it bores high into vertical banks. The nest cavity is lined with straw, grasses, and feathers, a major improvement over the cavity nesters in the last chapter. Some of these species—the rough-winged swallow, for instance—will only excavate if it cannot find a preëxisting cavity. Then there are the swallows that have relapsed wholly into passive reuse: tree swallows and purple martins rely on locating a cavity fashioned by another animal—such as the nearest nesting box put out by humans.

Shortly after the swallow family evolved, one new branch developed from a species that discovered how to employ excavated mud

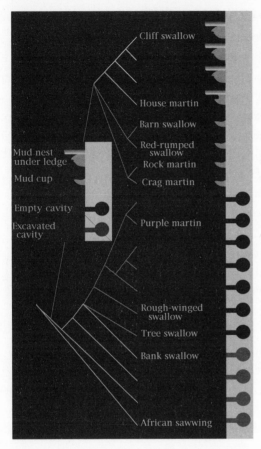

Nest evolution. A comparison of genetic sequences permits a reconstruction of the evolution of nest designs in swallows and martins. The descendents of the oldest surviving group excavate cavities, as do many closely related swallows (such as the bank swallow). But a behavioral mutation appearing first in tree swallows led to a group (which includes the purple martin) that simply renovates existing cavities. The other major group to evolve made use of a behavioral change that allowed them to build mud nests and cavities. This group includes the cliff swallow. Note that birds called swallows are not necessarily closely related; the same is true for martins.

to build an external cavity. The birds combined digging and piling of debris. From this point on, the strategy became one of harvesting mud pellets from the surface of damp ground, an excavation in miniature. The first use of this breakthrough was to build mud-based cups on vertical surfaces, generally under protective over-hangs. The barn swallow and crag martin still do just this, lining the cup with straw and loose feathers.

What is the behavioral or intellectual leap that allows a bird to see a burrow built out rather than dug in, with excavated pellets brought to the site rather than carried away? Could this be another example of behavioral duplication followed by an editing and re-ordering of the components, a kind of dyslexic exercise in building? Or does this require a network-mapping mind, one that is able to purposefully reorder elements and envision the outcome? In fact, a more dramatic cognitive threshold has been crossed: to building from the inside out, construction from the exterior has been added. The Tier–3 internal perspective (local-area mapping) has been turned inside out. The ability to imagine an object or structure from a new perspective is a classic test of intelligence, and it depends on Tier–5 mental manipulations. The need to externalize building required this step; the newfound ability opened the door for new uses of this cognitive tool.

The building behavior itself requires skill and determination. Compared to mere digging, attaching the first mouthfuls of mud to the cliff is quite a challenge. The bird chooses a spot on the basis of its inaccessibility to predators, and then faces the problems this very advantage causes. The bird often scrabbles desperately, using its tail not only for balance but as a source of resistance against the cliff as gravity inexorably pulls it down. Whatever the bird does, it must not yield to the temptation of grabbing hold of mud applied only that day, since it's sure to peel off. The animals typically build in the morning and feed the rest of the day, leaving their work to harden overnight.

The solution to the problem of attaching one pellet to an older one seems to involve two steps. The bird gathers a ball of mud of medium consistency, then smears a bit of soft mud across the top of its beak. Returning to the growing nest, it spreads the thin mud across the planned junction first, and then pushes the firmer pellet onto this surface. Next, the swallow vibrates its beak while still holding the pellet to encourage a finer joint; some researchers believe that this buzzing liquefies the mud on both sides of the joint, welding the new addition to the hardened foundation. (Some mud wasps do the same.) The cup nests that result are solid and elegant.

Cliff swallow. Cliff swallows build external cavities using hundreds of mud pellets. Though the general design of each nest is similar, the details of the execution vary according to the site and the individual builder.

The house martins, the next group to evolve, place their cups just below a horizontal outcropping, and build the side walls up to this natural roof. This architectural modification encloses the nest, adding still more predator protection and reducing exposure to the cold. The extra attachment surface more than makes up for the added weight. Again, construction efforts are divided between inside the nest and outside, requiring a constant shifting of perspective.

All three species of cliff swallows construct a curving entrance tube that points gently down. Doubtless this makes the task of would-be predators even more daunting; it certainly makes construction more time consuming, increasing the number of pellets needed from fewer than 1,500 to around 2,500. The external entrance tube mimics the entrance passage of some excavated nests, except of course that the structure is built out rather than dug in. The cognitive transformation of Tier–3 into Tier–5 mapping seems the most obvious way to enable this architectural finesse.

Though the nests of each of these species vary considerably (and often inexplicably) from one pair to another, and often show modification to match the contingencies of nest site and material, the basic design is species specific. With genetic analysis, the evolution of the building behavior is clear, and it follows the kind of progression we guessed at in the last chapter. What is missing from our picture is experimental manipulation of the nest. Researchers need to try adding or removing pellets, substituting materials, controlling the clay and sand content of the mud, reorienting design features, removing the overhanging ledge, and so on. Do the birds learn from mistakes and structural failures? In choosing the nest's attachment site, what factors do they consider, and how are they weighted? For birds so common, it's surprising how little the cognitive aspects of their building behavior have been examined.

Choughs and Ovenbirds

The invention of adobe cups that then develop into enclosed cavities, so evident in the progression from barn to cliff swallows, is also seen in some birds that build in trees. The magpie lark of Australia constructs a bowl of mud strengthened not only with grass but with small sticks, feathers, and fur. This creation is placed atop a branch, so it must be well secured. Not only is there less surface area between nest and substrate than the swallows and martins have to work with, but the branch can move abruptly, whereas cliff faces do not. One behavioral response to this greater challenge is that the birds seem to vibrate the mud more intensely, apparently to forge a better bond as it liquefies and runs between the straw and other reinforcing material. The birds also wrap the mud almost completely around the supporting branch, enlarging the area of contact. The magpie lark nest looks nothing like the nests of martins and swallows: the individual mud pellets cannot be distinguished, so complete has been the flow of liquefied mud.

Closely related to the magpie lark are two other Australian birds of the open forest, the apostle bird and white-winged chough. They build substantially larger nests, weighing up to five pounds, located as much as fifty feet above the ground. But even these scaled-up versions of the adobe cup with their inch-thick walls are manufactured with the same jiggled-mud strategy that seems to be universal among birds that build with wet earth. But then vibration is a key feature in the insertion of twigs and grasses into conventional nests, so this may be a bit of behavioral recycling.

The ovenbird takes the adobe cup to new heights, elaborating the simple mud cup into an enclosed sphere in the treetops. This group has about two hundred species in the New World tropics. Many fashion mud cups on branches, but six found mainly in the grassy plains of South America are classic oven builders. The pair waits for the rains, and then begins to gather mud. Like the choughs, they work plant matter in to add strength. In about two weeks these diminutive birds manage to work two thousand pellets, weighing about ten pounds in all, into an impressive dome.

The oven's construction involves building a rather ordinary but oversized adobe cup on the branch. This is then built up to make a sphere with a circular opening on one side, close to but not directly over the branch. Adding the mud pellets and smoothing them out without risking a collapse of the domed roof as it curves inward must require considerable care; the procedure employs behavior highly modified from that used for the cup. But the next step is in many ways more remarkable. The birds construct a curved internal wall about three-quarters of the way toward the roof, creating an entrance chamber between the door and the nest cavity. The indirect entryway becomes a severe obstacle for predators, and the smooth concrete-like dried mud itself repels attacks. The ovens also provide sheltered roosts for the family in bad weather.

That the ovenbird is something special is evident not just from the unusual nature and physical size of its nest. The enormous difficulty

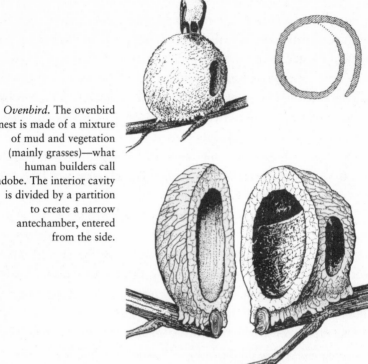

Ovenbird. The ovenbird nest is made of a mixture of mud and vegetation (mainly grasses)—what human builders call adobe. The interior cavity is divided by a partition to create a narrow antechamber, entered from the side.

of balancing the dampness of a pellet, its weight, and its placement relative to other pellets is daunting. As the chamber grows increasingly dark, looking for and counteracting sag in the dome must become very hard. But constructing the interior wall is equally difficult. It abuts dry adobe, and follows a three-dimensional plane that is not merely the outline of the bird or its foot-to-beak reach but a deliberate partition, unlike anything we've yet seen. What rules or design principles guide the builder? It must have a network-based Tier–5 map, even though it builds from the inside. The difficulty most humans have in interpreting a cross section of an ovenbird nest reminds us of the conceptual challenge involved. As with the martins and swifts, we know frustratingly little about a group that has much to tell us.

ENTANGLEMENTS

The "classic" bird's nest is based on the ancient platform design, formed usually from sticks or twigs. Material is added and worked from the inside to create a cup with a wall, which is then lined. Most of the building supplies come from plants of one sort or another. The wall helps insulate the brood, as well as protecting the eggs and chicks from accidentally falling or rolling out; the lining adds to that insulation, particularly under the eggs. The elements become entangled and thus stick together. In material-poor environments such as oceanic islands, birds such as the Peruvian booby, the Peruvian cormorant, and the brown pelican must fashion a substitute: they construct a nest wall of their guano, and then line the structure with their own molted feathers.

Other species in the pelican/booby and albatross groups, like their common ancestors, build nests of the same design from more conventional plant materials rather than from guano. Where grasses, twigs, plant down, stray feathers, and moss are to be had, bird nests can be (and generally are) considerably more luxurious. The ability to fabricate something of the same design out of guano, using quite different techniques, seems a reasonable instance of a goal-directed behavior; at the least, it indicates an innate picture and Tier–5 network-map flexibility since the building techniques (but not the nest shape) are entirely different. One scenario, in which the bird uses a mindless series of instructions in the manner of the funnel wasp, is absurd on the face of it. The more likely possibility is that a Tier–5 cognitive net allows the bird to select and optimize other elements of its behavioral repertoire to fashion its nest from lumps of damp guano rather than from sticks and grasses.

The familiar robin's nest is a good example of one variant of a widespread style of entanglement construction. The pair begins by placing or pushing sticks into the notch between the branches they have chosen as a nest site. Many fall to the ground on their own while others are dislodged as construction progresses; in time, a kind of mechanical natural selection leaves a roughly tangential collection of twigs that is held together by the natural friction of bark and rough surfaces as well as the entanglement of notches, which helps prevent sliding. The tangential arrangement maximizes the area of contact between the twigs and the consequent friction, and thus raises the odds that notches will become interlocked.

As its stability increases the birds begin taking the risk of landing on the platform and even doing a little bit of scrabbling—extending their legs backward to create a depression. As work progresses the stick lattice becomes more cup-like, molded as it is increasingly around the female's body. As she sits, she uses a quivering motion of her beak to insert finer twigs, a technique that allows each new element to find a path between and among the existing twigs. At least one layer of twigs is cemented into place with mud. This binder runs between the branches, gluing the structure together; some pairs continue this work on up, and actually use mud to form the rim of the cup.

Then the birds' attention shifts to insulation and lining. The insulation at the bottom of the cup can be moss, string, or grass. The lining is usually a circular array of fine grass, again held in place by entanglement and friction. Blades of dried grass are long, rough, and at once flexible and brittle. The bird pushes the thick end into the nest, and then sets about cracking the strand so that it runs around the edge of the inside of the nest in a series of straight or gently curving lines between the points of bending. By creating a stack of grass polygons, this approach leaves lots of insulating air between the loose-packed blades.

Robin nests are unmistakable, and indeed most species' nests can be identified almost at a glance. The set of common characters

Robin's nest. (A) Robins begin with a platform of sticks, then add twigs and mud to create a cup with high walls. The cup is lined with grasses, moss, and other insulating material. (B) In cross section, the layers of lining inside the mud-and-twig walls are visible.

that makes each species' nest distinct emphasizes the reality that most birds construct their nurseries on the basis of innate instructions. But equally obvious is the work of first-time nesters, whose structures are likely to be placed too low or a bit too far out on a flexible branch. These first efforts may be insufficiently supported (that is, not over the fork of a branch), or they may show evidence that the mud was too soft during building. Some adults, as we have said, construct high mud walls, but others leave the twigs of the upper lip of the cup exposed. One pair puts in hardly any grass; another packs in double the normal amount. Some of the variation is a matter of learning, some probably a reflection of genetic variation, but the rest appears to be what we might call individual preference. But what is the source of such preferences? Skeptics could assign it to sloppy programming (but then what has natural selection been doing with this critical behavior for the last few million years?); or it might be blamed on cognitive limitations—a sketchy local map of the actual goal. More likely, preferences are an indication that the birds rely on a network mapping of responses, with flexibility, learning, chance, and experimentation contributing to an optimization of nest building.

The general utility of the robin plan can be seen in a variety of other species. Rook nests, for instance, are constructed in a similar fashion, though they are substantially larger, and the rooks use no mud. The sticks in the rook's platform and outer cup average about eighteen inches versus four to six inches for robins; the thickest sticks, each about an inch in diameter, are on the bottom, in the crotch of a supporting limb. As the bird works away from the base, the sticks in this outer layer grow progressively smaller and thinner. The base of this deep cup is lined with a thick pad of moss and rootlets and their attached dry soil; the walls are lined with thin elastic twigs, inserted between one another, as we saw with the robins. Next comes a substantial layer of circularly arranged grass, and finally the job is finished with a full-cup cushion of pine needles when they are available.

Apart from a superficial difference (no mud), the concept is strikingly similar. Both rooks and robins build a platform base, then add to it to create an outer wall. Both bring material of ever-finer gauge. Both build their platforms by placing and poking, allowing friction and interlocking to provide stability. Both scrabble in the nest to shape the cup and fit it to their own dimensions, rotating from time to time to make a symmetrical job of it. Both work more and more from the inside (as opposed to atop) as the nest progresses, and both use vibrations to thread the material in. Moreover, once a new element is in place, the bird may pull back on it to be sure it has caught, and then reposition the twig if it is not secure. The birds compulsively seize loose ends projecting from the cup and poke them into an adjacent bit of cup, a tentative first step toward the weaving behavior we will be examining soon.

In some sense we are looking at similar examples of a wide and seemingly very different set of nests. Mallards, for instance, trample weeds into a platform rather than use twigs, and the final layer of lining is down, but the middle steps are very much alike, and the beginning and end serve the same ends. Heron nests are big on sticks and light on lining, and the final layer is leaves, but the concept and much of the behavior is similar. Kingbirds, phoebes, many flycatchers, most warblers (no sticks), cardinals, and sparrows (again, no sticks) follow the same pattern: the general order of ever-finer materials, a characteristic arrangement of platform elements, the raised wall, a scrabble-sculpted cup, vibration-enhanced insertion of materials, and a lining for insulation.

The platform-based cup nest, whether built on the ground or in bushes or trees, has a structural stability based largely on friction and entanglement. The shape is created by the bird, not dictated by a cavity, a declivity, or a scrape. There is a degree of discrimination, an ever-changing selection and workmanship not really possible in species that content themselves with tossing one or two types of material into a depression or onto a ledge.

If there is a classic bird's nest, the platform cup is it. But the minds, goals, and techniques involved have continued to evolve. Just as we saw with the mud-based structures, the brood may be safer if the nest can be enclosed. But enclosing an adobe nest involves something much more demanding than merely continuing the nest walls up and around: the roof is under a different set of stresses. New behavior, or behavior apparently based on some understanding of the problem, is

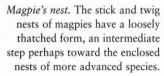

Magpie's nest. The stick and twig nests of magpies have a loosely thatched form, an intermediate step perhaps toward the enclosed nests of more advanced species.

necessary. Similarly, weaving overhead with twigs is an enormous challenge compared to building the cup. What can the structural elements be braced against?

The magpie solves this problem by continuing to exploit friction—the rough sticks they use provide just enough surface entanglement to hold together. They begin with the nest cup. The sticks of the platform and outer wall are notable for their many protruding side twigs, which create a rough and prickly appearance. The lining consists of a layer of mud, which seals the many openings between the sticks, and then a substantial padding of rootlets. Next, the bird builds a relatively open canopy of equally rough sticks, an arrangement not much like the cup's outer enclosure. The lack of density seems at first sight a result of carelessness or laziness, yet it not only keeps predators out effectively but also makes the actual entrance amid this seemingly random array of branches nearly impossible to find. But it's a difficult and expensive nest to build, requiring about seven weeks of the birds' time and energy to construct. The ground below is littered with material that didn't catch in the matrix.

Covered nests are also possible on a smaller scale if the surroundings provide some degree of stability—the kind of preëxisting three-dimensional platform that many spiders take advantage of. Thus leaf warblers, building in bushes, construct a conventional twig platform and cup and then continue it up and around, making use of the many branches and leaves available to support and stabilize the platform. As with the magpie's nest, and most of the other covered structures we will look at, the bird uses itself as the measuring instrument. For leaf warblers, the right distance from the cup to the roof is just what the bird can reach with twig in beak. The nest is necessarily spherical.

Carolina wrens do something quite similar. Though they prefer cavities, building in vegetation sufficiently dense that the bird can trample out a kind of cavity works well. They extend their cup of twigs above and around, using the surrounding herbs and grasses as a kind of lattice. The nest is lined and the entrance extended into a distinctive tube. And then there are the individuals that build the nest from grasses rather than twigs, and make it so structurally sound that it can be picked up and carried off intact. In many species of wren, the male builds several nest shells for the female to choose among; only the favored nest is ever finished, though one or more of the rejects may be used as a bad-weather roost. Since most birds that build conventional nests roost in the open, this custom suggests that the cavity design may provide extra benefits.

In general, however, a covered platform nest is difficult to execute. Imagine trying to carry sufficiently long and rough twigs (or even grasses) in through an increasingly narrow entrance, and then manipulating them from the inside to work them into the roof. And no two sites are the same; a single set of instructions would not work. This very difficulty seems to have selected for greater cognitive flexibility to meet the challenges, the kind we call Tier–5 network mapping, in which a multitude of behavioral options can be chosen among, ordered, and modified to achieve a goal. The flexible use of material and design concepts reaches new heights, and makes possible, at least in theory, novel design breakthroughs that could open the door to new niches.

Consider a wren that was building along the Delaware-Raritan Canal in Princeton a few years ago. This favorite spot for local anglers

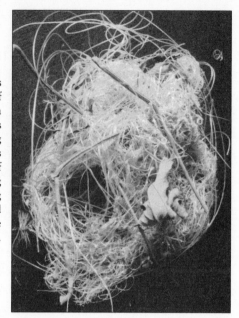

Fishing line nest. Carolina wrens vary greatly in the amount of building material they use in a nest. The more elaborate nests use great quantities of long plant fibers and grasses. This nest was constructed instead of numerous exceedingly long pieces of monofilament fishing line. More than a dozen lead sinkers and three hooks are incorporated into this structure.

is well supplied with monofilament line caught in the trees and bushes along the water. One male built a nest entirely of pieces of this fishing line, complete with lead sinkers and wire hooks—even a plastic float. Judging by the absence of lining in the nest cup, this strong and durable enclosed nest was apparently rejected by the female in favor of a less contemporary alternative. Had this novelty caught on, would a new behavioral subspecies have been born?

Despite the considerable advance in nestling safety and warmth the platform-cup nest conferred, the world of bird construction was waiting for another breakthrough—or several—that would free these brood chambers from the limitations of using twigs, or of having branches underneath, or of relying on mere friction. Let's look at how the various innovations, evolving first as apparently simple work-arounds, have blossomed and led to new levels of challenge and complexity.

STICKING TOGETHER

The key distinction in our summary of the platform–cup nest is that the bird searches and finds a site that provides a satisfactory set of supports; it does not *bring in* the supports. Another common thread in the platform–cup nest is that the animal creates a base, a bottom, and a side (and, in some species, a top). The platform and cup are mostly pieces of vegetation that are collected, placed, formed, and threaded into shape. Let's look at something completely different, yet fundamentally similar.

Saliva

The order of birds that includes swifts and hummingbirds split off and diversified after a single innovation: the development of hormone-controlled sticky saliva. Stickleback fish and bubble nest frogs also use secretions as a kind of construction cement. Many mud-building wasps, termites, and birds will add saliva if necessary to obtain the right consistency for their mud- or feces-based mortar. But the dense yellowish saliva of swifts and some hummingbirds is a tool of extraordinary versatility.

Neither hummingbirds nor swifts (commonly confused with swallows, to whom they are unrelated) find perching comfortable. Indeed, all eighty species of swift are almost helpless on the ground, and this has led to remarkable behaviors and even more fantastic legends. A common myth asserts that swifts spend the first three years of their lives flying, never landing even to sleep. Only the need to reproduce brings them down. The reality is that swifts roost in groups overnight, usually in hidden spots, and then only when darkness forces them to stop hunting insects; they leave that task to nighthawks and bats, who have the equipment to hunt in the dark. (Some seabirds, on the other hand, really do spend days or weeks on the wing. They allow one hemisphere of the brain to sleep while the other handles flying and feeding.)

Their aversion to perching is reflected in a common theme of hummingbird and swift behavior: these birds almost always collect nest material on the wing, and they build their nests while hovering. The most common swift in the temperate zone is the chimney swift, a species that constructs nests out of sticks. Before humans came on the scene, they built in caves or rocky recesses, but now they have chimneys. For the swifts, the first step is to create a secure foundation made out of their thick saliva.

Hovering within a chimney, they place two healthy deposits of saliva on the masonry wall one nest-width apart. Sticks that would be part of the platform of a conventional nest are then pushed into this sticky goo; the saliva eliminates the need for a supporting branch or shelf, thus opening up a host of new sites. The short sticks they use are all broken from dead branches while the birds are in flight. The first sticks are aimed down and across, and to them new saliva-coated twigs are added, so that soon a crescent-shaped outline of the bottom of the future cup appears. The nesting pair adds more saliva and new sticks to the original attachment points, angling them not only down along the line of the bottom of the cup but slightly away from the cliff. The growing nest is formed from a solid mass of twigs, loosely packed but glued firmly with saliva. The twigs of the completed nest cup glisten with a coating of saliva.

This is a strategy that need not be restricted to chimneys. Three small species from South and Southeast Asia (crested swifts, Indian tree swifts, and whiskered tree swifts) glue together larger pieces of material. They choose a likely spot high in a tree and then piece together a tiny nest less than an inch across on the side of a branch. Once the saliva has set, they lay an egg, which they incubate while standing on the branch facing in the direction of nest construction. Note that swifts' nests, being built from the out-

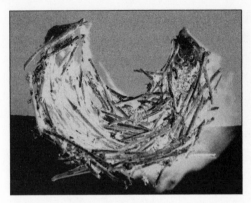

Chimney swift nest.
Chimney swifts apply two
large globs of sticky saliva to
a vertical surface, then begin
pressing in saliva-coated
twigs to create a cup-shaped
platform for their eggs.

side until the final step of lining the nest cup, require a cognitively
challenging Tier–5 network-mapping ability to maintain multiple
perspectives.

Chestnut-colored swifts build more substantial nests, using saliva
to hold together an otherwise impossible mixture of mulch and
moss. This species chooses quite wet sites, so that the moss continues to grow, its roots binding the nest to the adjacent soil. White-
throated swifts, in contrast, choose a dry rock crevice where they
glue plant down and feathers together. Such differences in preferred
nest habitat are essential if two otherwise-similar species are to coexist. Selection operates to increase the differences in nest-site preference, and thus the physical, behavioral, and neural specializations
these separate sorts of location entail.

Four species of palm swifts have discovered yet another way to
exploit the group's secret weapon. They catch feathers and bits of
grass, plant hairs and other fibers, snatching them out of the air as
they fly, and fashion a pouch by gluing the fibers to a palm leaf.
Long sticky strands of saliva are drawn across the back to hold the
palm blades together and reinforce the outer attachment of the nest.
The result is so tippy and subject to being whipped around, even inverted in the wind, that the swift takes the wise precaution of also
using saliva to glue her eggs to the nest.

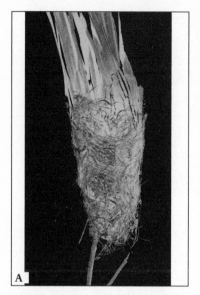

Palm swift nest. Palm swifts bind plant fibers together with saliva and glue them to a palm branch. (A) From the front, the result appears to be a deep cup woven to the palm, but from the back (B) it is clear that there is no weaving, and the attachment depends on thin strands of saliva.

A champion of the glue-nest builders is the lesser swallow-tailed swift of Central and South America, a relative of the palm swift. This bird spends the better part of six months gluing together a cylindrical creation of plant hairs, cotton fibers, and similar material that depends from an overhanging rock. Near the top is a chamber equipped with a nest cup where eggs and chicks alike are safe; below this is a remarkably long entrance tube so designed to make predator access difficult; a false chamber, tunnel, and entrance on the side a bit further up serve to frustrate even a well-motivated nest robber.

Selection has taken saliva building to its logical and seemingly absurd end with the subgroup commonly called swiftlets. These species live high in caves in Southeast Asia; a low-tech version of the echolocation by which bats hunt down insects allows them to find their way about in the dark. Some build strictly with saliva, but most species mix in feathers and some other apparently light-duty reinforcement. Still, it's mostly their personal brand of epoxy. The first layer, which forms the bottom of the nest cup, is applied (on the wing and in near darkness) along a downward-hanging semicircle. Subsequent layers begin just a bit above the original attachment point and angle out, creating a second semicircle in cross section. The resulting nest could hardly be better designed; each line of saliva dries so as to take the tension pressure between the cliff face and the nest rim. The vertical layers act against compression from weight in or on the nest.

The swiftlets take their fixation on saliva one step further. They embed the insects they catch for their young in a hard mucous matrix and serve them up as an insect-rich saliva cake. The nests, built entirely of saliva, are the key ingredient of bird's-nest soup, an expensive delicacy in many places. And it should be costly: swift nests, for their size, require a long time to build. However, there may be no other way to construct a sanctuary four hundred feet up in a dark, guano-filled cave. Saliva-based nests do not seem to provide much in the way of insulation, however, which leads to longer incubation times, so the evolutionary advantage does not lie along conventional lines.

The emergence of saliva as a binding agent has resulted in nests that look superficially like those of other species, but which require quite different neural equipment to construct. This is the first nest we have looked at that has to be built entirely from the outside; the guide to size as the nest grows cannot be the bird's own bulk. Any seamstress or model maker knows that interior and exterior work are confusingly different; switching from interior to exterior corner joints in finishing molding can leave even experienced carpenters with pieces an inch or two too long or short, and angled in the wrong direction. (As we have seen, many external-cavity builders start from the outside, but mix interior and exterior work as the structure grows.) What additional cognitive skills are necessary to think entirely outside of the nest? Without even being able to perch, with the bird's body axis in constant flux relative to a fixed cliff or leaf, surely some concept of what their final product should be is essential.

Cables and Silk Framing

Like the swifts, hummingbirds hover when they feed, and many species use the same thick quick-drying saliva in nest construction. But though they catch insects to feed their young, the adults' main food is nectar. And their nests are made primarily of vegetation rather than twigs. The ruby-throated hummingbird, the only species found in the northeastern United States, starts its nest with a firm disk of saliva. But a less obvious behavioral quirk that makes hummingbirds quite different from swifts lies in their use of an invertebrate secretion as well: silk.

Silk, as we saw, is a strong but flexible material, and the spiders' webs that multiply as the summer progresses provide ample resources. Except for the catching threads, most spider silk is free of adhesive. We have noted that many webs are composed entirely of threads used to knock down flying insects, platforms to trip them up, support threads, and refuge tunnels. Hummingbirds find a likely nesting spot and then begin bringing in spider silk. Because it is for the most part not sticky but easily tangled, getting the silk to adhere to twigs and rough bits of bark in the vicinity is a matter of looping the strands around projections a few times and then tugging. The pressure of the latest loops on the earlier ones does the trick. Collecting and building both occur during flight.

When the hummingbird has made a start, it begins alternating vegetation and silk. The building materials depend on the species of hummingbird and include the finest, most insulating moss and plant down available. The work might begin with some plant hairs, secured to a branch by aerial wrapping with silk. If the hairs are long enough, they, too, may be wrapped, and the ends pushed through the growing mass on the branch. New bits of material may be flown in and threaded through the upper periphery of the base, and then pulled back down; after several such additions, the bird wraps another cable of silk around. Other construction items—bits of bark, lichen, moss, or the plant wool that helps seeds disperse on the air—may be pushed into the structure, a process called *felting* by some researchers.

Other species of hummers work at forks in branches, and they use the silk first to form a hammock upon which to build. With the addition of material the base sags, allowing some of the original silk to serve as a framework for the nest wall. The sagging is essential for producing a nest large enough to accommodate the eggs and an incubating parent. The bird attaches a cable, circles the nest a few times, and then adds more material. The new silk cannot usually be wrapped around a branch; it consists of many strands, and is simply pushed through the tightly packed vegetation already in place. This process creates an expanded knot of silk on the other side, which thus anchors itself to the growing nest. These birds all work their nests from both sides, integrating interior and exterior perspectives and techniques.

Many hammock nests are not built by hummingbirds at all. The nest of the pied monarch, a flycatcher living in the rainforests of Australia, joins two adjacent vertical twigs, to which the cup twigs seem to adhere by magic. In fact, the base is a silken hammock on which a twig has been placed. The bird ties this first twig with loops of silk to one of the verticals and then repeats the process, tying twig after twig either to the support or to another twig, or to both. Finally, the bird lines its nest and adds a bit of lichen. The cup nests of vireos look similar at first glance, but have been firmly lashed to adjacent branches with silk cording at every step of the way. The interesting thing about the hammock nests of hummingbirds and other species is that the beginning looks and feels nothing like the end product; at least with cup nests, the platform is where the nest will be, and the growing cup encloses the exact space and shape that will become the nest.

Hummingbird nest. This seemingly shapeless mound of debris is actually a tiny cup nest made of moss and lichen woven together with spider silk.

The way nests are finished is also sufficiently characteristic to allow many hummingbird and other silk-based nests to be identified at a glance. Birds may attach lichen to their nests with spiderweb; they may line them with moss; some use silk to anchor their nests to nearby twigs, and some weave a silken rim around the nest's edge. Hermit hummingbirds often construct nests that hang from a branch. The support links the tree to only one side of the nest, creating an obvious problem when the parents start to incubate; the nest will rotate until the center of gravity (the bird) is directly below the support, with the inevitable loss of all concerned. Hermits respond to this challenge by building a long counterweight under the center of the nest, which helps prevent tipping. Some species add substantial quantities of dry clay or pebbles, all held by spiders' webs, to lower the center of gravity yet further and thereby reduce the tipping when the nest is occupied.

Hummingbirds, as we have noted, alternate between working from inside the nest and building on the outside. When inside, their tiny feet and weak legs are useless for the scrabble-sculpting of conventional birds. Instead, they use their beaks to place material and push it through. They continue to fuss with the lining while incubating. At first glance, their nest building looks pretty flexible. But is it?

Observations of what Konrad Lorenz would call a natural experiment tell us that once a hummingbird is sitting on its nest, the drive to brood slows down or shuts off some kinds of cognitive processing. H. O. Wagner observed some white-eared hummingbirds nesting in Mexico. Theft of nest material is fairly common during the building phase, but one pair had the misfortune of nesting near a compulsive robber. The female was sitting on her nest with two hatched young while a violet-eared hummingbird made occasional passes, plucking out bits of vegetation from the wall. During construction, hummingbirds will repel thieves and generally repair any damage, but in an impressive demonstration of situational intelligence—or lack of it—the brooding bird allowed the robber to keep picking away at the nest for a week, by which time both chicks had fallen out.

We presume that hummingbirds have at least some kind of Tier–3 local map of what they are trying to build, and a Tier–5 network-map ability to orchestrate a set of innate motor programs and recognition systems and deal with the repeated switches between inside and out. Their nest-building behavior is just too flexible and adjustable to unpredictable local contingencies to seem less cognitively complex. But once the birds segue from building to incubating, it's as though they have undergone a lobotomy. The drive to stay on the nest, to remain inconspicuous, overwhelms any thought of the larger issues. With so much ordinary behavior specific to particular contexts, the idea of context-specific intelligence seems hardly surprising.

Hummingbirds, as we have seen, are by no means the only kinds of birds to have discovered the utility of spider webs. Probably the most interesting use of silk's ability to be stretched without breaking is seen in bush tits, long-tailed tits, firecrests, goldcrests, and a few other remarkable species. These begin with a loose hammock of silk hanging from a twig frame. To this they add a seemingly random collection of plant wool, moss, and lichen. Then one of the birds settles into the hammock and, by virtue of its weight and some careful scrabbling, stretches the structure into a cup. More vegetation and silk is added, and additional bouts of internal molding cause the structure to sag into a deep pouch. When the nest is deep enough, some species even build a roof and an entrance, before adding a final round of silk and decoration. Each of these hanging nests is luxuriantly fitted out with a generous lining of feathers.

There is good reason to think that the choice of materials is not random. Mike Hansell has studied the nests of bushtits and long-tailed tits under high magnification. Not only was he able to confirm that no weaving is involved, but, more important, that the choice of lichen for all but the finish coat is specific. The birds always employ a species that has long blades ending in a multitude of wiry rhizomes. Whenever the bird works to expand the nest, the silk slips past these springy grasping fingers, only to be caught again later when the pressure slackens. The bird is ratcheting the fabric strands past the framework, and it adds more of each as the cup thins from the stretching. Hansell rightly points out that this kind of attachment is analogous to the working of hook-and-loop fasteners; the rhizoids act as the rough hooks and the silk is equivalent to the smooth loops.

This stretch-and-lock strategy appears again in conjunction with a much more impressive strategy for building a framework: weaving. It's difficult to believe that the Velcro-using birds have much idea of what they are doing at the level of hooks and loops. More likely they have a sense of what feels right at each stage. Again these are species that must work from both sides of the fabric, keeping track of where the bag is getting thin and thus needs reinforcement. Some sort of concept of their goal seems essential, and so does a neural net to choose and orchestrate the behavioral elements needed to complete the project. Designing a feathered robot able to deal with unlikely contingencies would be far more complex. But the secret of the cognitive underpinnings of silk-based nests lies in the carefully conceived and controlled manipulation of nests and available construction materials—tests yet to be performed.

Sewing

Silk's strength, flexibility, and seamless length make it a nearly ideal thread for sewing, and the long, thin beaks of some birds immediately suggest needles. But only two species, so far as is known, take advantage of this synergy. One is the long-billed spider hunter of Southeast Asia. This nest is built on the underside of a leaf, typically of a banana plant. The bird punches holes in the leaf about three inches to each side of the strong midrib; at first there are just a few, but as the work progresses the number rises to about eighty on each side. The bird collects long strands of spider silk and pushes an end through one of the holes. A brief tug tells it whether the tangle of silk on the other side of the perforation is going to hold. Then the bird flies to a hole on the other side of the midrib at about the same height and pushes another silken wad though. In this way, a simple and loose sling hammock is created, which the bird gradually fills with long plant fibers.

As construction continues, the width of the sling narrows to close the bottom of the nest, and new lines are added to reinforce the structure. Eventually, a silken cable runs across the width of the leaf every quarter of an inch or so. More and more fibers are brought in, and soon the builder is pushing in between the plant material and the leaf to press the cup into shape. In the end, the hammock is about sixteen inches long and six inches across; its interior width is two and a half inches, its depth four inches.

At least as impressive is the work of the tailorbird of India and Southeast Asia. The astounding tailorbird's nest consists of two leaves sewn together with a thick mass of silk. The bird punches the

Tailorbird. The bird finds two likely leaves, punches a series of holes, and then uses spider silk to pull first one pair of holes together, and then another. After this framework is ready, the bird fills the cup with low-density material, such as moss or fungi.

holes first—many more than will be needed, but then some later tear under the pressure they must bear. The two leaves are pulled close together, and one end of the clump of silk is pushed through a hole in the first leaf, and then through a corresponding hole in the other. Something like two dozen stitches generally suffice. The resulting pocket is then stuffed with moss, lichens, fine grass, plant wool, or some other suitably soft substance.

Because these birds are common in populated areas of India, any number of natural experiments have been observed. Unlike those involving the hummingbird whose nest was picked apart, these observations are rather to the credit of the tailors. For one thing, if the leaf chosen is large enough, the tailorbird will fold it back on itself and stitch the seam; this requires about half the work of a two-leaf nest, the leaf itself sealing the other side.

Tailorbirds seem to have some more complete understanding of their task—an understanding that allows them to exploit shortcuts and substitute string or long flexible plant fibers (or even thin strips of bark) for silk in case of need. The argument for some greater comprehension of the goal on their part, some ability to use and even modify the innate steps to take advantage of fortuitous contingencies, seems fairly strong. There is at least a local-area picture of the nest, and probably a cognitive map for the home range, a cognitive network to account for the flexibility, and possibly even a Tier–6 concept map. But the silk-sewing strategy itself has not led to much speciation; that is, it has not opened up new niches, but rather given a couple of species a better way of exploiting their existing spot in the fabric of nature.

WEAVING

In the quest to create a platform, a basis for attaching and building a nest, we have seen birds move away from scrapes in the unforgiving ground, and even from natural cavities affording more protection but few creature comforts, to ever more elaborate structures. These nests are in general safer and warmer, but they also require more physical labor, and perhaps greater intellectual ability as well. We have seen the development of rock, stick, and mud platforms, the invention of actively excavated and externally constructed cavities, layered nest cups on arboreal platforms, saliva platforms, and the use of silk and saliva as cup- and platform-binding agents. Weaving and threading and even sewing of nest cup material has been evolved to increase stability and structure.

The strategy of using silk to bind the first elements of platforms to branches, and then later elements to each other, is a huge breakthrough. Indeed, in these nests, the platform is really the outer edge of the nest cup; there is no need to wedge coarse elements into the crotches of trees or low bushes. But the downside of tying the nest cup to a branch with silk is simple: for all its remarkable properties, silk is not strong enough for most nests. It may not break easily, but it does stretch, and stretch, and stretch; enough weight, and the parents, brood, and nest will hang down too loosely from the supporting limb. Some wholly different materials are needed for larger birds to exploit trees fully.

"Incidental" Weaving

We have seen birds working with many kinds of flexible materials, including grasses, plant fibers, roots, snakeskin, and anything else strong and pliable. The key to a strong functional nest is to make a stout initial connection to the branch or other support. Woven connections provide a stable flexibility ideal for conditions where support structures are in pretty constant motion. This is perhaps easiest to imagine for birds that live in reeds, a really excellent habitat. There is water below to discourage many predators, seed heads above to provide cover from aerial hunters, and a nearly endless supply of emerging insects to feed on. Red-winged blackbirds and reed warblers, to mention two, use reed grasses for the outer cup. These are woven in and around several nearby stalks, with an occasional loop. The grasses are sufficiently rough and pliant that they tend to stay put long enough for more to be added, thus locking each other in place. Soon there is a vague cup that can be elaborated and lined with finer material.

Whether or not this strategy relies on weaving is debatable, as experts disagree on what should qualify as "true" weaving. Anyone who has made cloth will find the technique of blackbirds laughable; it's nothing at all like our kind of weaving in which threads or other elements are worked in and out of another array of threads or supports in a systematic way, and doubled back repeatedly for strength. The blackbirds seem to be haphazardly winding grass stems, and the stability of the result is a surprise. It depends on friction and a sort of structural inertia. Still, to the nonweaver, it's easy to consider the result as woven because long thin elements are wound through other long thin structures until the binding of the upper part of the cup, and perhaps the entire fabric of the nest, is strong and fairly rigid.

The most difficult of nests to build (at least in the human imagination) are probably the hanging pouches of orioles, penduline tits, and the cup-and-tube nests of weaverbirds. The most important challenge for these birds is to create an attachment to the supporting branch, and then a work platform. The birds bring long thin material—plant fibers, adventitious roots, or grass—and attempt to loop it around and around the branch. Tits and orioles make no attempt to tie a knot: They hold down the irregular spiral they have made and poke a new piece of material into it, pull it through, and loop it around the branch. Sometimes by apparent chance they double the new end back before threading it into the growing mass again; this fortuitous maneuver makes a knot. The birds seem to have no concept of a knot as such, or at least have great difficulty executing one. Try tying a knot with a grass blade and a pair of tweezers and judge for yourself whether the problem might just be mechanical.

The weave is extremely irregular, but more and more fibers are looped into the existing structure, lengthening it down from the branches. Soon the two sides of the hanger are joined; orioles usually fuse several hanging strands. The birds continue to insert long fibers and pull them through, spotting loose ends and pushing them in again and again until they disappear into the structure. By now the bird has fashioned a loose and rather ratty fabric into which it inserts filler—the so-called felting. For orioles, this is almost always downy plant seeds, rootlets, and short bits of vegetation, so that the side begins to look finished even though most of the material is held only by friction, and easily picked out.

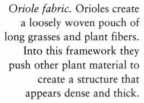

Oriole fabric. Orioles create a loosely woven pouch of long grasses and plant fibers. Into this framework they push other plant material to create a structure that appears dense and thick.

The nature of the supports dictates the location of the entrance. Since orioles have multiple suspension points, the opening is at the top surrounded by the hangers; but because the tits have only one support, the nest is built with gaps at the top fore and aft. As the nest nears completion, the tits seal one opening off and finish the other as a short, neat tube. The African penduline or kapok tit builds two nest cups, one above the other, each with its own entrance. The upper one, which leads to the brood, can be folded against the rest of the structure where it is almost invisible; predators able to find the nest at all usually examine the empty chamber and then leave. Regardless of species, the nest involves at least ten thousand insertions of material; presumably the benefit of living in a hanging nest a hundred feet high is worth the staggering effort involved. But even then the job is not finished: The nest must be meticulously lined. Only after weeks of effort is it finally ready for the eggs.

True Weaving

Most researchers consider the weaverbirds, a family that can be found in Africa, India, and Southeast Asia, to have the most intriguing nests on the planet. Weaverbirds work with thin strips far longer than those used by tits and orioles, and they employ a fundamentally different strategy. They begin in the same way, by tying a knot. This process is harder, because the material is long and unwieldy, but also easier, because after weaverbirds pull an end through they are very likely to reverse direction and push the point back in the opposite direction. In short, they seem at some level to understand knots, at least as experienced adults. Another apparent breakthrough: Weaverbirds employ green vegetation, which is more pliable, easier to weave, and can readily be knotted. Some weavers use long grass; other species tear long strips from a leaf, generally incorporating a strong rib; some strip out thick plant fibers.

But the weaving behavior is what is so obviously different. When two parallel strands are in place, a weaverbird is likely to wind a blade of grass or plant fiber down under the first, then back over it and down under the second, then back over it and down under the first, again and again until there is nothing left to work with; then it spots the other end of the fiber and does the same with it. No filler is needed; bare spots are simply targets for more strands. The vibration of the beak as the new end goes in is the same as in most birds, but the compulsion to seize, draw out, reverse, and plunge the end back in is as striking as it is effective.

Once he has tied the initial knot, the male begins to weave a loop to perch on. Using himself as the ruler, he builds at the limit his neck and beak will reach, weaving strand after strand in and out of the loop with few reversals, a process often called twining. The perch can be an inch thick when it is finished. Often he builds the loop from just above a downward branching point of two twigs, making the initial knot above the inverted Y so that slipping off is unlikely.

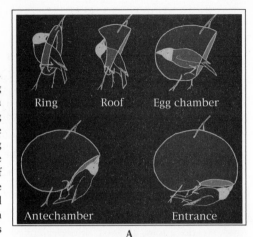

A

Weaverbird building.
(A) The first step in building for a weaverbird is tying a knot where the top of the ring will eventually be. Next, the bird creates the strong ring upon which it will stand while constructing most of the rest of the nest. He then weaves the roof (which will be thatched later) and egg chamber (which will be lined if a female accepts the nest). He next builds the antechamber and the entrance hole. The entrance tube is added only after the male has a mate. (B) The ring (shown here with the roof) is extremely thick. This is the only part of the nest a male will not repair.

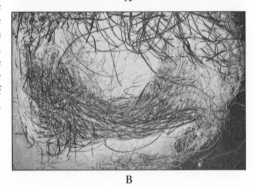

B

Now the bird weaves out at beak length to create the roof and sides of the nest, and continues on around to generate the nest cup. The cup at this point is relatively thin, and of a surprisingly regular rectilinear weave. Both roof and floor still need a lining, but of different kinds. The object of the roof lining is to keep the brood dry. For weavers working in grass, it's time to switch to wider blades; those weaving fibers must pack in layers and layers of new material. For both, the work is more a matter of thatching than weaving: no reversals, no knots, and mostly long stitches. The floor of the nest, which will later be given layers of softer lining by a female, needs little additional work at this point. We should keep in mind that the

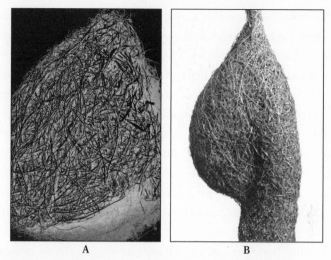

Indian weaverbird nest. (A) A weaverbird's nest before the entrance tube is added, and (B) afterward. The nests of this species have pointed roofs.

male works from both the inside and the outside of the structure, and thus has to switch perspectives frequently.

Then without turning around the bird starts a curved half dome which he builds down to the horizontal. Some species simply build a reinforced circular entrance at this point, but most leave it unfinished in anticipation of adding an entrance tube. In either case the male now begins displaying in an effort to attract a mate. He hangs upside down, flaps his wings, and calls desperately. Any interested female will inspect the nest, tugging at the walls, looking up though the roof for any sign of light, and turning about in the cup. If she accepts the male, he begins building a long entrance tunnel to discourage predators while she lines the cup by inserting soft material into the woven matrix.

Thanks to the patient studies of Nicholas and Elsie Collias, among others (and the remarkable willingness of weaverbirds to breed in temperate-zone aviaries), many of the questions that come quickly to mind about weaverbird behavior have been put to the test. For instance, experienced males seem to know that green is the right color for nesting material, selecting it more than 60 percent of the time (versus the 17 percent expected by chance) when five other colors of exactly the same material were available. But this is an inborn preference that also matures. On the first day, naïve males chose green about 40 percent of the time, but by the sixth day it was their first choice almost 80 percent of the time. Something about working with green strands just feels right.

Learning the best length of material, on the other hand, seems to require more experience. First-year males gathered several hundred grass blades with an average length of less than eight inches, and 16 percent were a useless two or three inches long. Second-year males, on the other hand, brought in strands averaging eleven inches, including nearly half that were a full fourteen inches, and just 1.5 percent were in the pathetic two-to-three-inch category.

The speed of weaving and the tightness of the finished product are also related to experience. Though we must not forget that birds build these complex and unlikely nests in the same general way on the first try, the initial nest is a loosely woven, shabby affair. Moreover, although the males prefer green vegetation, it naturally follows that a nest built slowly, as first tries invariably are, will be brown by the time it is complete; the deft work of experienced birds, on the other hand, will turn out nests that are partially or largely still green.

Charles Darwin noticed that most mate-choice decisions are based on contests between males—competitions for desirable territories with an abundance of resources, for instance, or a position in a hierarchy that might depend on physical stamina, coloration, or size. But in some more recently evolved species, females choose males for less tangible reasons. Peahens, for instance, select among peacocks on the basis of the number of ocelli (eyespots) on their tail feathers, the symmetry of the tails themselves, and the vigor with which the males wave and shimmer them at passing females. After mating with the male of her choice, the peahen builds her own nest and rears the young as a single mother.

The logic of female-choice sexual selection is to enable the female to locate the male with the best genes by using some set of correlated characteristics that do not depend on blood tests or interviewing family and friends. Weaverbird nests have the potential to be just such a genetic test. For example, female weaverbirds prefer green nests over brown ones. Take two mostly green nests with displaying males, paint one nest completely green and the other brown, and the female will choose the green alternative.

Females also take into account the vigor of the male's display and the brightness of his coloration, both of which are indirect measures of his health, and thus his genetic resistance to the parasites and diseases of the area. And, as we have implied, the quality of his nest material and the general workmanship of the nest are critical.

What this tells us about the cognitive processing of females is not entirely obvious. The color patterns and striking display ceremony of the males are doubtless sign stimuli that communicate species, sex, and reproductive readiness. The vigor of the performance seems designed to excite the super-normal stimulus circuits in the female brain. But where does the preference for tightly built green nests come from? Can these criteria have evolved out of nowhere, or are the females applying some sort of vaguely sensible analysis? They have many nests to compare; what attracts them to the newest and most solidly constructed? Why not just pick the largest? And given that all weaverbird nests will turn brown long before the eggs hatch, why is brown inherently wrong during courtship?

Nicholas and Elsie Collias were also able to look at the way the birds control building and nest repair, and thus to uncover the operation of a robust decision-making network even in a species that works exclusively from the inside. One striking set of tests involved hacking off great pieces of nests that were under construction. Rather than giving up on his race against the clock, the male would

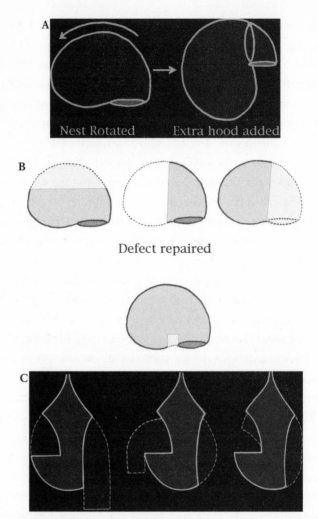

Weaverbird tests. (A) The bird continues weaving the antechamber until the entrance is horizontal. As a result, it is possible to rotate the nest and force the male to continue building. (B) Weaverbirds are able to repair a variety of damage, major and minor, including the complete removal of the roof, nest chamber, or antechamber. However, severing the loop, even after the rest of the nest is complete, leads to the destruction of the nest. (C) Cutting out only a section of the nest cup before work begins on the antechamber presents males with an apparent ambiguity. Some simply rebuild normally, others repair the undamaged antechamber to create a nest cup and build the entrance on the opposite side, and still others seem to try to make two nest cups but give up in seeming confusion.

reweave the roof, the entrance, or the egg chamber. What he would not do, and probably cannot do, is repair the bottom of his initial loop, the scaffolding on which he must stand to build. If the perch is cut away, he will tear down the entire structure and start over.

The ability to repair a structure such as this is cognitively demanding even with just a local-area map of the structure, but another researcher found that individual weaverbirds have a more flexible picture of the goal. Thus males can come up with strikingly different solutions to the same damage. Excising a wedge that includes the lower part of the roof prompts some birds to replace the missing part and then move on to build the entrance. Others, however, opt to rework the side with the large opening, turning it into the wall of the nest; the damaged area is reworked into an entrance. Others start along this path, but misjudge the angle of the new entrance and so abandon the renovations. Still others botch the project altogether, enclosing both openings.

The Colliases were also interested in what cues terminated building. As in the funnel wasps, the angle of the entrance opening controlled how much roof was to be built; they wove until the angle was right, and rotating the nest prolonged the effort. By placing green felt over the egg-chamber roof, they were able to make the males stop thatching. When they replaced the egg-chamber roof with screening, they found that the males went ahead and thatched this metal fabric if the mesh was fairly open, but they were unable to make anything of fine screening. In the second scenario they just moved on to other parts of their construction project.

Finally, the researchers explored the decision to stop lengthening the entrance tunnel. They found that although the male had a strong sense of where it should stop, an opinion that differs from one bird to another, they could trick him into adding more by the simple expedient of threading another strip into the tip of the tunnel. The male, extremely fussy about loose ends, would weave and weave until the two new ends were buried. Adding another blade forced him to lengthen the extension further; he couldn't bring himself just to pull the strip out and drop it. In the extreme, this tactic drives a male to extend his antipredator structure all the way to the ground, defeating its purpose.

Weaverbirds know more about what they are building than our funnel wasps; they can respond sensibly to unlikely damage, except the destruction of the scaffold. They can go back to a step they have completed and redo it; if a piece of felt induces them to stop thatching the roof and move on to the next step in construction, they will return to thatching when the felt is removed. Their consternation at the appearance of a fine mesh screen is both understandable and baffling. If the male cannot physically fit the grass blades through the openings, it makes sense to give up trying; on the other hand, if the purpose of thatching is to keep the brood dry, why doesn't he tear the nest down and start over? And trying to put a neat finish on the entrance tube, no matter how long it takes, represents the most inefficient kind of rule-bound behavior.

Male and female birds of many species appear to operate with at least a partial grasp of what they are about as well as a strong drive to achieve their goals with the innate and learned behavioral tools at hand. The most intriguing aspects of their behavior come from those areas where they depart from their innate instruction set and do something that seems genuinely creative, or even aesthetic. Larks, for instance, sometimes add a small inexplicable ramp of pebbles to one side of the nest; rock wrens, one species of phoebe, and a wheatear do the same. Rock nuthatches, which build the largest external cavities known, often press colorful beetle wings into the soft mud before it dries. And don't forget the almost pathological fads and preferences of penguins for special stones.

On what is perhaps a related note, some pairs in conventional cup nest species add a partial or complete roof, others a far higher wall, yet others a superabundance of lining. Bird architecture exhibits an individual variability that in humans we would call aesthetic. What is its source in birds (and humans)? Are their cognitive preferences and tastes akin to our own? Or perhaps, after all, a subtle element of context-specific survival value, which we are too dim to pick up on, could underlie some of our own supposedly objective and logical cognitive judgments.

The architectural creations of one group of birds have nothing whatsoever to do with nesting. Thus, with practical and utilitarian considerations largely irrelevant, the role of aesthetics, individual variation, and female choice take on added weight. It is to the almost whimsical bowerbirds and the minds that create their remarkable mate-attraction edifices that we turn next.

CHAPTER 8

Bowers

OF ALL THE STRUCTURES animals create, none are stranger or more wonderful than the mating and display stages that male bowerbirds construct with elaborate care. Thatched with sticks, clothed in mosses and ferns, decorated with meticulously arranged flowers, stones, insect shells, or feathers, perhaps even painted with berry juice, the bowers may come to more than fifty times the weight of the bird that fashions them.

The bowers are not nests. Females build their own brooding accommodations in trees, and they rear their offspring without help. The bowers are, instead, critical mate-attraction sculptures, painstakingly evaluated by females—and by competing males, ready at any opportunity to vandalize the competition. And so they are just as important for the reproductive success of a male as more conventional avian creations.

And yet, if any aspect of human behavior can be divorced from utilitarian purpose—if, for instance, art may be said to be simply a useless sensory or neurological extravagance, a pleasure-inducing stimulus whose roots in natural selection are obscure at best—then the ontogeny, individual variation, construction methods, and preferences associated with bowers may be very close to crossing some important line between the mental experiences of humans and other animals.

Bowerbirds can indulge in this frivolity, if that is what it is, because there are few predators and little competition for food in their habitat much of the year, and many species have a very long or continuous breeding season—factors that have allowed artistic creativity to flourish among humans as well. Perhaps this aspect of the bowers' creative uselessness, this apparently excessive focus on the aesthetics of construction, helps explain the affinity many commentators from Darwin on have felt with these outwardly absurd birds. It may also account for why Western researchers believed for decades that bowers must be the work of diminutive undiscovered forest tribes.

At the cognitive level, these creations are the most complex seen in birds. Only beavers and humans undertake work with more steps and greater flexibility in design, materials, and execution. Bowers require a juggling of multiple perspectives and parameters. And because males are competing desperately to reproduce, with only this artifact to save them from genetic oblivion, selection has strongly rewarded the sorts of mental tools we generally associate with intelligence.

THE SATIN BOWERBIRD

The bowerbird family consists of seventeen species, all restricted to Australia and New Guinea. Different species are found in mountains versus plains, tropical forests versus near desert, mangrove forests versus temperate grasslands and woods. Together, the range of bowerbirds encompasses about two thirds of the area of Australia and New Guinea. A typical bowerbird looks something like a small crow (though rarely black), about ten inches long including tail, and sports a fairly heavy-duty bill.

The two most ancient members of the group, the spotted catbird and green catbird (no relation to the temperate zone catbird), are monogamous; the pair builds a nest and rears the young together. This is the pattern in about 90 percent of birds in general; more often than not, the males can do more to advance their reproductive interests by building, incubating, guarding, and feeding their off-spring than they could by abandoning the female and seeking additional mates. But already among the catbirds we see a curious and seemingly inexplicable investment in unnecessary building: The pair maintains a rudimentary circular display area of upside-down leaves. As soon as the leaves wither, the birds replace them. What end could this effort possibly serve? Perhaps it's a now-meaningless inheritance from an unknown ur-bowerbird.

The common ancestor probably did construct arenas of some sort. According to DNA-sequence analyses, the nearest relatives of bowerbirds are the lyrebirds, a family of only two species that live in Australian forests. Male lyrebirds create a display court by piling up a mound of dirt; they dance on this stage, utter an amazing range of calls they have learned to mimic (including, alas, the sound of chainsaws), and show off their two striking tail feathers. The mounds, however, are not decorated; this elaboration seems to originate with bowerbirds. Thus the catbirds' display circle is in part a puzzle, an artifact in search of a purpose. We may be looking at a pointless *jeu d'esprit* that became the basis for the bowers of the other members of the family, or the degenerate expression of a display-area building behavior that evolved early but withered away as the catbird genus relapsed into conventional monogamy. Or perhaps we just do not yet know enough about these two species.

The rest of the bowerbirds are roughly divided into two groups. One set of species builds avenue bowers, the other fabricates what are called "maypole" bowers, though this term hardly describes the range of variation. We are going to look first at the satin bower-bird, an avenue builder, simply because it is by far the best understood. A native of the moist forests of Australia, the satin bowerbirds' habitat stretches 1,800 miles along the southeast coast. The males

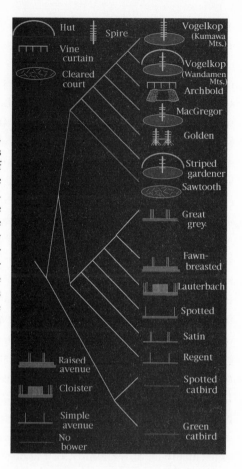

The Bowerbird phylogeny. Genetic sequence analysis yields this pattern of relationships among the various species of bowerbirds. Australian catbirds are the modern representatives of the oldest group, most closely related to lyrebirds. The other two general groups build either avenues or maypoles. (The sawtooth bowerbird has apparently regressed to a style similar to that of catbirds.)

regularly build their bowers near human settlements where researchers have the advantage of roads and electricity, something that cannot be said for the remote highlands of New Guinea or the bush of Australia.

The satin's relatively modest bower begins with the choice of a suitable site—one that is open and flat and where abundant light penetrates the canopy. The male starts by clearing away any remaining leaves that seriously shade his arena, sometimes even defoliating small trees. Next, he removes all the debris from an area about a yard square and then disposes of the useless leaves and slivers of bark some distance away.

Now he brings in hundreds of sticks and twigs, which he drops and tramples with seeming casualness where the bower and display platform will later be built. The sticks generally fall in a roughly circumferential pattern, just as we saw with the stick platforms of many nests. This apparently offhand effort results in a structure with the same surprisingly solid friction- and interlocking-notch-based strength of more conventional passerine nests. The thick firmly interlocked platform winds up strong enough to be picked up and rotated (by researchers), or even carried away. Presumably this behavior derives from the first stages of nest fabrication.

The next steps in the satin's avenue construction seem at first glance like nothing we see in nest building, though attempts have been made to find links. Indeed, if there is no evolutionary continuum, how did bowers originate? Surely what appears to be totally novel—a behavioral special creation—is the result of a long and convoluted series of evolutionary modifications. Surely. The male satin bowerbird begins collecting twigs from eight to twelve inches long, which he places vertically in a double row running north-south; nest-building birds never insert twigs vertically. When there is a curve in the twig, the bird often takes advantage of this nonlinearity to create a bit of an arch in and over the avenue that is developing between the lines. Again, hundreds of sticks may be involved, and the result is two walls each about as thick as the male himself.

Researchers who assume that the bower results from a redirection of nest-building behavior see this as the nest-lining phase, but the conclusion is strained. It may well serve as a screen for the female to hide behind while observing the male's display; the bower owner is extremely possessive, and likely to attack the very females he is working so hard to attract. There is considerable variation among males as to how long the avenue should be (between fifteen and twenty-five inches), how densely to pack in the sticks, whether an

Satin bowerbird building.
The simple avenue of the satin
bowerbird begins with a dense
mat of twigs into which two
rows of vertical sticks are
placed to create the bower.

interleaving of straw is acceptable, exactly how close to north–south
the avenue must be (none more than 30 degrees off have ever been
observed), and the size of the display platform at the north end.

More work follows, and with it even more individual variation.
Some birds add finer twigs and even grass to the display arena
where the male will sing his lyrebird-like songs and court passing females. A satin bowerbird may paint the interior of the bower avenue with the juice of crushed berries that he holds in his beak, or
saliva-covered pieces of charcoal, or chewed vegetation. Some birds
have even discovered that they can achieve better results by using a
piece of fibrous bark as a brush. The effect, as judged by humans,
can be spectacular. Again, there is nothing remotely like this in cup-
nesting birds, or in the entire animal kingdom for that matter.

Next, the male gathers objects that he places about the north
end—the sunny part of structures in the Southern Hemisphere—but
their choice and placement is at once consistent and idiosyncratic.
This is no randomly selected collection of decorations. The items
that males of this species pick are highly biased toward dark blue
and purple, colors almost nonexistent among the local flowers,
feathers, and insects. Quantitatively, were the male choosing flowers randomly, there would be 95 percent fewer blue and purple objects. But on the other hand, this is no mere perverse preference for
rarity: Red, pink, and orange blossoms are also very hard to find,

but they *never* turn up as ornaments. (Yellow and white objects are used, but with restraint.)

The dark blue and purple ornaments, particularly the much-prized parrot feathers that males fight over, match their glossy bluish purple plumage; females, on the other hand, are a pale yellow green. The decorations also include some bright yellow leaves, and the displaying male is likely to hold one of them in his beak, which is also bright yellow. To a student of sexual selection, the leaf display looks like a way of producing an artificially enhanced beak as a low-cost supernormal stimulus, and the dark blue-purple decorations appear to visually enlarge the male himself. Others argue that the decorations represent an evolutionary relic from an ancient custom in which males courted by feeding berries to the females, who perhaps had some special preference for bluish fruits. This somewhat unlikely hypothesis reminds us of female balloonflies, whose cognitive capacity for distinguishing between substance and show is probably more limited than that of female bowerbirds.

Whatever the logic, blue certainly matters to males and females alike. At the first hint of wilting or desiccation, the bower owners replace those hard-to-find blossoms and berries. Though they build their bowers well apart, males regularly raid the competition, carrying off blue ornaments; for good measure, they pull sticks from the walls of competing bowers. In habitats that provide a wealth of decorations, the raids focus almost entirely on vandalism.

By their differential ability to build, decorate, steal, and destroy, then, the males in an area set up a social hierarchy based on who can steal the most decorations while losing the fewest. In one study, the top male carried out twenty-five successful attacks on his colleagues while suffering only eight on his bower; at the other end of the scale was an individual whose six raids against others were made at the cost of thirty-one against his own structure. But what really matters here? Males who originally built large, symmetrical, dense bowers did not have any particular advantage at the search-and-destroy

Immature bower. Young male bowerbirds produce loose, poorly constructed bowers. This is an early attempt by a satin bowerbird.

game; instead, those who were assertive raiders—generally the oldest and healthiest among a species that has a life span in excess of twenty years—wound up with the most decorations and the least structural damage to their bowers.

What does this mean for reproduction? Of thirty-three males in one region, the five most successful achieved 56 percent of the matings. Obviously, the system is highly biased, and the females are responsible. The hens are in no way coerced to a bower; even the top-ranked male must display and wait, nervously rearranging twigs in the walls. When a female comes, she usually just watches and leaves. Male dominance plays no direct role at this stage, but female choice among those useless bowers and the males that attend them *is* highly correlated with the quality of construction and the number of preferred decorations, demonstrating that accommodating female preferences pays. Of course, as we saw, bower quality is as much a matter of being able to protect your own construction and decimate the competition as it is of building an outstanding one in the first place.

Male satin bowerbirds seem to require a long apprenticeship to be competitive in the marketplace. Early building attempts may come to nothing more than a small, loosely woven platform to which a few vertical twigs have been added. Several youngsters may coöperate on these trials, but even such pathetic efforts are destroyed by older males. Some immatures attempt to paint the interior of a bower (often that of a mature male who happens to be away) without first crushing their berries or adding saliva to the vegetation. At this stage, the young males still have female coloration, and their incompetence could be entirely a matter of age. Many innate behaviors, such as flying, mature, transforming from an uncoördinated set of ill-timed and disorganized actions into a perfect performance in a matter of days or weeks, often without benefit of experience.

Thus, depending on how the behavior unfolds, the building practice and visits to the bowers of high-ranking males could be critical, helpful, or irrelevant. The variation in design—size, density, finish of the display platform, and organization of the decoration array—could be genetic, learned, or creative. One piece of evidence that may be relevant is that when testosterone is implanted in four-year-old males, inducing in them the plumage of seven-year-old reproductively active birds, their building skills do *not* improve. (The treated males are also harassed by their elders, who are neither fooled nor amused.)

Female choice based on nest quality is not unique—we saw that with weaverbirds. An unreasonable indulgence in ornamentation is also common among species with female choice, but is most often expressed in response to coloration, plumage, the number and elaborateness of songs, and the frequency of displays. The usual explanation for such preferences when the male can provide nothing of material use is that natural selection has tuned the female nervous system to recognize good genes in males—the most she can hope to

get out of the brief partnership. Of course, her system for recognizing males of her species can also be tricked with extra stimuli of the
right sort—for example, more ocelli in the tail if she is a pea hen.
But what about all that variation among bowers? Is there a variation in female preference that maintains it? And if so, what is the
source of these differences in taste between females? Alternatively,
are males experimenting or expressing competing drives with different mixes of artifacts and design features?

In short, though we know more about satin bowerbirds than any
other species of bowerbird, we cannot say with certainty what is
happening cognitively. We do know that males bring internal and
external perspectives to bear, benefit from learning, and (most important) they can usually repair damage to their bowers. When repair is more trouble than starting over, the male will disassemble it
and begin again. This means that the builders have some understanding of the overall design; these behaviors cannot result from a
rote building routine. But how did what understanding there is in
bower building develop? Did it happen over generations, or does
every generation learn from experience and observation? How much
does this understanding have to do with nest construction? What do
males and females comprehend about bowers? Are preferences for
structure and ornaments genetic or cultural? Is there some sort of
aesthetic process at work? Oddly enough, it is among the species we
know less about that many of these answers, or at least strong hints,
are to be found. It's not that a lack of facts is actually helpful, but
rather that a few bits of comparative data about other species can be
used to evaluate what are otherwise glib and untestable hypotheses.

OTHER AVENUE BUILDERS

The three simplest avenue bowers are the products of the evolutionarily oldest avenue-building species, the regent, satin, and spotted

bowerbirds. The regent appears more primitive, the spotted more advanced. The regents throw up simple stick avenues about seven inches long on small but sturdy platforms; they add an average of ten rather mundane decorations such as green leaves, brown fruit, and some snail shells. The avenues, such as they are, do at least get a coat of vegetable paint. Courtship begins in the forest canopy, and the male leads the female to his bower. Females do not seem to care a great deal about the number of ornaments, but males who have built the better bowers are more likely to mate with the birds they have attracted for a closer look.

Spotted bowerbirds, which thrive in unpromising habitats of coarse grasses and scrub, are the most widely distributed members of the family. Instead of needing to seek a well-illumined hole in the forest canopy, spotted bowerbirds are on the lookout for shade-giving bushes, generally near water. Some reports describe east–west bowers, but others find no consistent orientation; males display in the morning, so an east-west orientation would make sense. The avenue itself is constructed of straw, fine twigs, and grassy stems, and is generally larger and more substantial than the satin bowerbird's. The walls range from ten to twenty inches high, the avenues from six to nine inches wide, and the overall length is from fifteen to thirty inches. The variation between bowers is also more extreme.

The spotted's bowers are decorated with stream-worn gray-to-white pebbles, white snail shells, and up to a thousand sun-bleached bones, usually from rabbits. In their quest to supplement what nature provides, the males are famous for looting human habitations for eyeglasses, car keys, and anything else white or silver, shiny, and portable. Like the satin bowerbird, this species paints its avenues. Mating is highly skewed: three males in a region populated with thirteen bowers accumulated 60 percent of the copulations. Compared to satin bowerbirds, however, the spotted conducts few raids either to acquire ornaments or to inflict damage; perhaps this is because the arenas are so few and far between in this poorer habitat.

Nonetheless, it would seem that such raids would be useful if they were practicable as female choice in this species is strongly correlated with the number of decorations and the quality of the bower.

Taking these first three species together, we see tremendous variation and one consistent correlation: The avenues can be north–south or east–west; raids can be critical or unimportant; decorations can matter or not; the construction can be heavy duty or lightweight; but no matter what, the females demand that the bowers be well built and durable.

We can look at the other three more recently evolved species of avenue-building bowerbirds about which something is known to see whether even this one correlation survives. All three—Lauterbach's, the fawn-breasted, and the great grey—have heavy-duty platforms. Indeed, Lauterbach's and the fawn-breasted bower bases are almost absurdly thick, and even the great grey's base is 50 percent thicker than that of the first three species we looked at. The great grey's lifestyle is similar to that of the spotted bowerbird's, except that it lives in tropical areas of northern Australia where from twenty to sixty inches of rain falls every year, whereas the spotted's ability to survive on five inches allows it to subsist over a larger range to the south. The grey's bower has the usual two rows of sticks, to which are added twigs and grasses. The avenue is from eighteen to thirty-six inches long and from fourteen to sixteen inches high; the walls are six inches thick. Located in the shade of a tree or bush, the great grey's bower is oriented north–south.

Debate continues about whether the avenue is painted; though the males have been seen going through the motions, the color of the walls appears unaltered. If this is painting behavior, it must be pretty mindless, but we may be falling into that easy trap of assuming that animals sense only what we find obvious and important. Birds can see in the ultraviolet, for instance. If a component in the birds' saliva reflects this wavelength, and the color it reflects is attractive to females, the behavior could be critical. In addition, there could be an odor attractant in the saliva that humans can't detect; one powerful paint additive might be its perfume.

The great grey's bower is unusual in that both ends are decorated, though only one is used for display. And the decorations are remarkable: Hundreds of sun-bleached kangaroo bones and white shells are set against green ornaments, including berries, seed pods, leaves, and flowers. Added to these are opportunistic decorations such as bottle tops, metal buttons, nails, silverware, keys, and other human detritus. The ornaments are "zoned"—that is, they are arranged into groups of similar color, continually rearranged by the male. No very extensive experimental work has been done on this species, but we are probably safe in guessing that females prefer bowers that exhibit impressive collections of white objects set off by things green and sparkly.

The fawn-breasted bowerbird occupies low-lying, often coastal habitats. The very real possibility of flooding probably selects for the thick platform, which can be as much as fourteen inches high. Into this substantial foundation the male erects a twig avenue perhaps a foot tall and one to two feet long, aligned from east to west. He paints it green, and may weave fine rootlets along the top of each wall. Finally, he strews green berries liberally at the west end of the platform, inside the avenue, and along the tops of the walls. For the first three hours after sunrise he displays in front of the decorated west end.

The most remarkable avenue builder is the Lauterbach's, also known as the yellow-breasted bowerbird. This overachiever constructs an inner bower that is framed by outer walls at right angles; as a result, the inner runway faces into the walls of the outer avenue. This cloister design is cited as evidence by researchers who maintain that the structures serve to protect females from the males' aggressive behavior. The inner avenue is from seven to thirteen inches high, and of fairly conventional design: a thousand sticks and an equal number of grass stems. But the outer parentheses are another matter. This is

Lauterbach bower design. Lauterbach males basically build two avenues, one inside the other, at right angles. The inner wall of the outer bower is decorated with many pounds of stones.

the only avenue in the family that leans out. Each of these secondary walls is fashioned from another thousand sticks, between which the bird inserts approximately five hundred slate-gray stones at each end with the precision of a stonemason.

The Lauterbach's decorations are quite variable; he makes use of bluish gray berries and pebbles, and red berries and fruits. Typically, from 5 to 10 pebbles are used, but one bower had 130; the number of red decorations ranges from one to three dozen. The bower platforms must support all those sticks and stones, and are thus quite thick and well constructed. The Lauterbach's bower, which weighs in at as much as sixteen pounds, can be lifted and carried away without structural damage. The males, for some sense of scale, weigh about four ounces, a sixty-four-fold difference. In human terms, a male weighing 175 pounds would need to fabricate a courtship structure of about 11,000 pounds to impress a female with this turn of mind. (Impressive as the Lauterbach's sculpture is, the bowers of other species can weigh hundreds of pounds.) Given the remote habitat of this species, the experimental questions are unanswered. What does seem obvious is that females require males to make stupendous efforts and to employ skills that extend to carpentry, masonry, and interior decoration.

The consistent correlation among the species with avenue bowers is that females demand and recognize high-quality work. Quite often they insist on a great deal of building, and it must be nicely decorated as well. Every component within a species shows variation, but enough consistency prevails over a sufficiently wide range of locales for us to be fairly sure that the basic design and color choices are innate. In fact, a consistent similarity persists between body color, beak color, and the color of the crest behind his head that the male flashes at the female during courtship on the one hand, and the colors of the decorations on the other. One hypothesis suggests that elaborate bowers evolved to substitute for gaudy colors: Both serve as potential signs of male genetic quality, but fancy plumage imposes a more direct predation risk.

The data from the satin bowerbird strongly suggests that learning and practice are crucial to successful building. Their facility at repair tells us that the males can picture the end product. The idiosyncratic edifices and decorating schemes of different males and their constant fussing to try new variants (usually just to return to the original arrangement) implies an element of something like individual style. And that we find these improbable collections of sticks and other debris so appealing must make us wonder whether there might not be some kind of common animal aesthetic at work. This last somewhat unsettling thought is given far more ammunition by that other group of bowerbirds: the so-called maypole builders.

MAYPOLE BOWERS

The most ancient extant member of the subfamily of bowerbirds that are considered maypole builders does not, in fact, build a maypole. The sawbilled or stage-maker bowerbird has a remarkable notched beak that the male uses to flatten the large leaves in his display arena. The site is in jungle or forest, and begins with an oval stage from three to eight feet in diameter, swept clean of debris. He covers this arena with about forty leaves, almost always upside down, which he replaces at the first sign of wilting. (Plenty of variance is evident; some arenas have a hundred leaves.) If the bower is decorated, few ornaments are involved; no one has observed any being brought in, and the few that have been noticed may have been in place before the male began to build. The cleared, carpeted, circular arena is an almost constant feature in the members of this group, but the stage itself bears a strong resemblance to those of the distantly related monogamous catbirds described earlier.

Archbold's bowerbird, which lives at still higher elevations in the New Guinea mountains north of the sawbill's range, looks at first glance to be on the evolutionary high road to maypoles. In one locale where a few arenas have been found, there is an avenue from three to eight feet long flanked by two berms of dead fern fronds. On these parallel bunkers are arranged discrete piles of snail shells and one of gaudy beetle elytra (wing covers). One bower had, in all, 135 brownish shells, some of which appear to have been painted blue on the inside, and twenty-seven elytra.

On the leafless low branches of adjacent trees the male had draped dramatic yellow vines (climbing bamboo), which he replaced as they wilted. Also on these branches, as well as on nearby fallen trees, were bits of charcoal, berries, and more snail shells. To add to the complexity, though, Archbold males in a different mountain region build walls of interlaced twigs eighteen inches high and about thirty inches apart in a cleared oval court, which they decorate at each end with charcoal and black berries. These bowers have neither ferns nor hanging vines.

Accounting for this remarkable variation along conventional lines of thought is a challenge. In terms of evolution, we would have to suppose that the two groups have been reproductively isolated long enough for their arena-building programs to generate two unique strategies—strategies as different as in the least-related species in the family—but that at the same time selection has left the birds themselves morphologically indistinguishable. Another alternative is even more unlikely, but probably more nearly correct: the arena differences may be culturally transmitted. On an island with so many isolated mountains and valleys, tradition can be very local indeed, as evidenced by the myriad of human languages in New Guinea. We've seen enormous variance in the building of satin bowerbirds, as well as evidence that learning has come to play a major role in their behavior, and this in a species that nowhere suffers from isolation from others of its kind.

MacGregor's bowerbird is the quintessential maypole maker. The male MacGregor's selects a three-to-six-foot sapling in his dense rainforest habitat and defoliates it; around it, he clears a circular arena four feet in diameter and carpets this area with a thick layer of moss. One bower, built on a slight slope, had a level moss platform ten inches thick on the downhill side. The vegetation is usually thickest next to the sapling and at the edge of the circle, creating a circular avenue from two to six inches thick around the maypole. In building the platform the bird collects distinctively colored moss from high in the trees, ignoring ordinary ground-growing moss. So thoroughly intermeshed is the moss that it can be cut from pole to periphery and rolled up like a carpet.

Around the sapling the male works hundreds of twigs ranging from perhaps four to twelve inches, with an equivalent variance in diameter. Reversing the conventional large-to-small sequence of cup nesters he uses the smallest twigs first, laying them horizontally, tangential to the sapling. As with the sticks in conventional nests and platforms, friction and interlocking notches hold these elements in place. The moss piled up at the base of the pole provides additional initial support, as do the side branches of the sapling over its entire height. As the male works higher and higher, the sticks he

MacGregor's bower. Like most of the maypole builders, the MacGregor's male begins by creating a thick, circular layer of moss. The maypole itself is enveloped in sticks and decorated with ornaments.

uses are thicker and longer, giving the maypole its distinctive shape. To build up to six times their own height, birds must be able to perch on these heavy-duty sticks and the sapling's branches. When researchers disassembled one fairly average bower, they found it comprised 816 sticks and twigs, nearly all gathered from near the top of the seventy-foot forest canopy; a tall bower would require at least twice this number. In design and execution, there is nothing remotely like this among nest builders.

Although the male MacGregor's does not collect ornaments for display on the ground, essentially all the maypoles have sticks decorated with white fungi or tassels of insect or spider silk, generally on twigs near the ground. Either the males are constantly redecorating or they are raiding each other's poles; sticks there one day are often gone the next, only to be reinstalled later. Neither raiding nor redecoration has been directly observed, though vandalism seems the most likely explanation. There appear to be major differences in the degree of ornamentation in different regions, but whether they are genetic or the result of cultural learning is unknown, as is the effect on the choice behavior of females.

Some male MacGregor's bowerbirds begin a second or even third maypole near the primary one, though in the end the twigs in them are removed and worked into the central pole. But the bower of the golden (or golden-fronted) bowerbird requires two poles from the beginning. The male must find two likely saplings from three to eight feet high that are growing from three to five feet apart. Though they tend to be similar, there can be as much as a twofold difference in height. Critical to the edifice is a singing perch that can be anywhere between one and five feet off the ground. This can be a natural branch extending from one tree to near the other, or a fallen branch that has become entangled in the two trees, or vines linking the two poles together.

Unlike the MacGregor's, which begin with small twigs at the bottom, golden bowerbirds use their largest material first. Working the sticks in horizontally, the male slowly creates two cone-shaped maypoles. As more sticks are added from top to bottom, the outer edge begins to sag and hang, concealing the horizontal scaffolding within. The cones grow until their bottoms overlap, creating an elevated runway hill between the two poles (and under the singing perch) about a foot high. He carpets the runway not with moss but with lichens, and decorates the sides of the two cones that face the runway with fresh olive-green flowers or with dried creamy flowers that have shiny black seeds attached. Some bowers have pale moss, ferns, and berries near the base of the display stick; others may have white moss and clusters of grape-like berries lining their avenues.

Bear in mind that in this leech-infested jungle, fewer than two dozen bowers have ever been found; building has never been seen in progress, and even courtship remains to be observed. Ornament theft has been documented, but how much females care about the decorations, much less the quality and symmetry of the building, is anyone's guess. Based on what we know of other species, though, it's likely that the females value at least some characteristics of the bowers very highly indeed, which is why males invest so much time in constructing them. But the one thing that is most evident from the data is that variation between bower building and ornamentation is enormous, and the source of this diversity is unknown.

The streaked bowerbird (also known as striped or orange-crested) erects a fascinating elaboration on the maypole theme. The sapling and its hundreds of interlocked sticks and circular moss-covered platform are there, but two new features make this structure probably the most beautiful of all animal creations. First, the same woven black moss (mixed with black fibers from tree ferns) that the bird uses for the platform is hung in a continuous sheet up at least the outward-facing side of the maypole to create a dark vertical cylinder nearly a foot in diameter and perhaps two feet high. Into this moss, like a jeweler placing precious stones on black velvet to enhance the contrast, the male inserts most of his decorations. In a typical bower, scores of iridescent-blue beetle wings on one side and shiny snail shells on the other set off a central vertical line of yellow flowers. Florist as well as decorator, the male replaces the flowers as soon as they wither.

The builders take considerable care over the placement of these decorations, arranging them so they are not too crowded. One observer describes how a bird returning from a collecting expedition studies the array anew, places one new-found decoration and hops back, tilting his head, to take another look; then he either moves an existing ornament to a new spot or takes a newly gathered one in his beak to install.

The second remarkable elaboration is the hut the male builds over his display cylinder. This he fabricates out of hundreds of interlocking twigs so tightly and thickly worked that it is largely waterproof. It is a kind of display case for the maypole, serving to keep the decorations safe from all but the worst rains; it also may provide a place for the male to retreat out of sight of the wary female attracted by his calls while she examines his bower. The window analogy is apt because the twig work continues down to the moss platform and then encircles it on the open side, covering its two to four inch thickness and adding an elevated entry rim. This threshold is itself heavily decorated with fresh berries, generally bright red. Reports of striped bowerbird display arenas are unfortunately so rare that nothing definitive can be said about variation, ontogeny, or female choice. A closely related species, however, has proven more amenable to observation and experimentation.

The Vogelkop (or brown) bowerbird is far more common, or at least more accessible to the determined researcher studying them, than the striped bowerbird. Classic descriptions of its arenas depict a moss-mat platform encircling a maypole. The pole itself includes hundreds of horizontal twigs, and both mat and pole are encased in a tightly worked hut between two and six feet in diameter and from eighteen to thirty inches high. Unlike the striped bowerbird, the Vogelkop uses green moss, and he does not cover his pole with it. There is, however, a moss cone about a foot in diameter and six inches high at the base, and the platform extends about a foot and a half out the door. Instead of putting the ornaments on a column of moss, the builder places his decorations in distinct piles both inside and outside the door of the bower.

Perhaps the most consistent theme in the study of Vogelkop ornaments is the wild variety of male decorating decisions. Many males, for instance, will collect dramatic black bracket fungi and arrange them in a pile roughly six inches high and eighteen inches across, somewhere between three and fifteen feet in front of the door. Other males will make two piles, some will collect none at all, and a rare individualist might put a pile inside the bower itself. Here and there, a bower owner will collect orange bracket fungi instead.

Vogelkop hut bower. In some parts of its range, the Vogelkop bowerbird (often called the brown gardener bowerbird) constructs a tightly thatched hut around its maypole, and decorates its moss mat with piles of carefully sorted ornaments.

Beetle elytra are also a common feature, but whether there are four or thirty-two or some number in between depends on the male, who piles them either in the hut or on the welcome mat in front of the door. The same pattern holds for blue fruits. Some males also add piles of orange fruits (more often outside than in), red leaves (always outside), and so on. Another observer working in 1872 reported that the vogue was for piles of green fruits, flowers with rosy apple-like berries, fungi, and "mottled insects" completing the decorating scheme. In 1938, the bowers seemed to feature primarily large brown decayed fruits and huge flowers. An arena recorded in 1944 had piles of fresh flowers, yellow fruit, mushrooms, charcoal, and black stones. In 1964, a group of nine bowers included ones favoring blacks, browns, and blues, but other designs focused on red, blue, and green. There can be hundreds of bright flowers, or none; a bower owner may collect and display seven hundred large blue fruits, or four small ones, or none at all.

In most bowers of this species the display objects were grouped by type *and* by color—orange berries in one pile, orange fungi in another—but in a few, only color seemed to matter. The key word here is "matter": the males are not just making mistakes or being careless, as can be shown by moving decorations by as little as two inches while the bird is away. The owner notices the change, often immediately. Sometimes he restores the misplaced object to its former pile; otherwise he carries it off and discards it. Decorations are brought in or shifted daily; a pair of butterfly wings may be added to the outside mat one day, moved indoors the next, redeployed a few inches the following day, and then discarded. When researchers offered a choice of objects to the birds, not one opted for white berries, and only one chose acorns. This variation is not related to hypothetical differences in the local availability of materials; when poker chips of seven colors were set out near bowers, males collected some colors preferentially and piled them up in mounds. Blue was generally popular and white ignored, but males differed in their preferences. Each bird gathered its favorite colors day after day, and stole the same colors from rivals.

These variations are like nothing seen in the mate-attraction and building behaviors of other animals, and they provide much food for thought. But before we try to interpret the differences in construction and decoration in this species, we must consider a different population of Vogelkop bowerbird, first discovered on the Kumawa Mountains of New Guinea in 1981, and now known to be quite common. Just seventy miles from one concentration of hut builders is a large population that omits the hut. The maypole is much taller (up to nine feet) and includes hundreds of sticks from eight to thirty-six inches long, *glued* to the sapling, apparently with saliva; no other bowerbird is known to use glue, though fungal growth helps cement some maypole elements together in the towers of golden and MacGregor's bowerbirds.

Although the circular moss platform and cone around the base of the pole is found in this population, it is woven from fine dry fibers

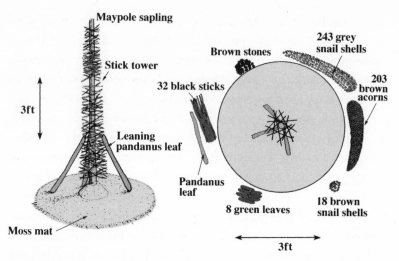

Vogelkop pole bower. In some areas, Vogelkop males elaborate the maypole and omit the hut. The decorations are shown in the top view (right).

of dead moss, and then painted black. The source of the pigment is apparently the birds' own excrement, which is unique in the area for not being the typical avian white. The platform mats are nearly perfectly circular, and range from four to six feet in diameter; the platform itself is about five inches thick, with neither raised rim nor depressed alley. The moss platforms alone weigh up to an astonishing 235 pounds; no one knows how long they take to build.

Nearly all bowers studied involved large pandanus (screw pine) leaves, which are about twice as long as the birds and half as heavy. Some (three to six per bower) were leant against the maypole, the rest arranged in a pile just beyond the edge of the mat. As with the other decorations, no obvious relationship is apparent between pole height, mat diameter, and number of ornaments; each male seems to be judging the optimum relative investment separately, and coming up with different answers.

Bowers in this region have hundreds of gray or brown snail shells, either one color or the other, or both arranged in separate piles or lines. These shells are not easily come by: The scientists studying the birds were able to locate only about two shells an hour while looking for nothing else. Scarcity is not the main attractant, though; massive piles of acorns, which are very common in the forest, are also popular. Bundles of thirty to ninety sturdy sticks (painted black) were also a regular feature, along with 10, 50, 150, or hundreds of dark brown stones, objects the researchers could not find *anywhere* in the habitat. One male also piled up seventy of the abundant white stones. Beetle wing covers, so popular among the hut-building population, were less in demand here: One male gathered forty-four wings, another five, a third only two, and others omitted this element altogether. One bower had eight green leaves, another two brown leaves and eight black fruits, another five brown leaves, but a fourth had neither leaves nor fruit. And what happens with poker chips? In this population they are ignored when left several feet away, but rapidly removed and discarded if placed at the edge of the mat.

Though the Vogelkop is extreme, we've seen lots of individual variation in bower size and decoration in other species. Bower design itself is sometimes the focus of innovation in other species. In studying great bowerbirds, Clifford and Dawn Firth found an attractive structure that differed from the usual plan. Great bowerbird males normally construct an avenue bounded by two gently curving thickets of sticks. (The sticks themselves curve inward overhead to partially enclose the bower.) One male added a pair of flanking thickets, also curved (right-hand drawing):

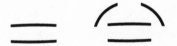

We see at least a superficial resemblance to the structure built by Lauterbach's bowerbird, a species in the same genus. Is this novel design an aborted behavioral duplication or a genuine innovation? And if females prefer the new design, could we be seeing the first steps of speciation? Could female choice, genetics, and learning be working in concert to bring about change?

Based on what we know about the genetic variation of behavior in other animals, it's safe to say that some—probably most—of the differences between male preferences seen in bowerbirds are not encoded in the DNA. The extreme variation also seems to make any model based on rote learning unlikely; memorizing, on the whole, leads to greater similarity. The change over time that seems to emerge from comparing observations over the last 125 years suggests that there could be fads, since males, despite their differences, may focus one decade on flowers and another on fruits. The extraordinary variation in a single habitat could be interpreted as selection for novelty—a drive to come up with something new, a familiar motivation in mate competition in our own species. At the same time, the huge differences in bower design and decoration strategy in the same species in different places has the feel of cultural tradition, or perhaps insight and innovation. In short, much about bowerbird architecture implies cognitive processes more like our own than those required for even the most challenging nest building. The most extreme interpretation of this family's abilities, however, goes even further: Could bowerbird females and/or males be driven by a sense of aesthetics, a concept of "beauty"?

AESTHETICS?

Darwin confronted the question of bowers in the larger context of sexual selection. He needed to account for traits, including behavior, that did not appear to help an animal survive. The bright colors and awkward feathers of peacocks, for instance, have no role in prey capture or predator defense; indeed, they attract the attention of predators and slow escape, and the females have no such burden.

Instead, he argued, the behavior and morphology seen in (usually) the males of certain species was entirely a matter of propaganda. And for such advertising to work, at least the females must have a general sense of what is pleasing to the eye.

As Darwin put it:

Sense of Beauty.—This sense has been declared to be particular to man. . . . [W]ith cultivated men such sensations are intimately associated with complex ideas and trains of thought. When we behold a male bird elaborately displaying his graceful plumes or splendid colours before the female, whilst other birds, not thus decorated, make no such display, it is impossible to doubt that she admires the beauty of her male partner. . . . [T]he playing passages of bowerbirds are tastefully ornamented with gaily coloured objects; and this shews that they must receive some kind of pleasure from the sight of such things.

A hard-nosed skeptic may object that, for all the individual preferences and regional fads, the overall species-specific nature of bower design and decoration points to innate circuitry. Where there are multiple designs and strategies, they might represent alternative programs to be used as circumstances dictate. To a skeptic, any sense of beauty or aesthetic delight is an illusion, an artifact of the neural wiring that controls the release of pleasure-inducing chemicals in the brain—a consequence of natural selection's favoring individuals who have certain preferences.

Perhaps. But then, what would lead us to suppose that some facets of the human sense of beauty are not innate? Is our nervous system immune to selection? For instance, newborns just old enough to test prefer the same photographs of female faces that college students find most attractive. They find bright colors more interesting than dull ones, and they focus special attention on shiny and glittering objects.

Social Intelligence in Animals

Social 0:	Social isolation: conspecifics are either ignored or attacked.
Social 1:	Social hierarchy: animals have a linear representation of part or all of the social order, especially individuals ranked near the individual in question.
Social 2:	Decision-network mapping: multidimensional representation of parameters is important in making social choices.
Social 3:	Attribution and intention: animal has an ability to understand the cognitive processing in the brain of a conspecific, and can alter its behavior to exploit that knowledge.

Of course, we have a lifetime to cultivate finer discriminations and unusual tastes, and a language to use in the process, but bowerbirds must create any counterculture tendencies in only a few years. But can we say that the human aesthetic is entirely different in origin and development from that of the one family that builds structures just for display? Beauty for bowerbirds and humans is probably such a prominent part of behavior because such a large proportion of our reproductive success depends on it. Predation is not a major threat, and finding enough food is not of overweening concern for many individuals for much of the year; thus, there is plenty of time to obsess on attractiveness.

Bowers are the ultimate example of architectural show, unconstrained by the need for conventional utility. These sculptures are never really finished, being constantly renovated and "detailed." The degree of external perspective required, especially in managing the decorations, is unique among birds. Only a concept of beauty accounts for this seemingly *de novo* behavior, based apparently on scores of simple unrelated innate and learned motor programs, recruited or perfected to execute the birds' "vision."

Another part of the male behavior may involve an ability to appreciate the female's perspective—a controversial Social–3 level of interaction known among philosophers and primate researchers as "attribution." Some social monkeys and apes, in at least certain contexts, behave as though they understand the intentions of others,

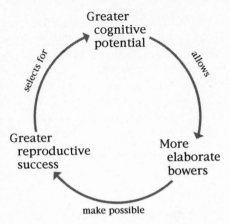

Positive-feedback loop. The positive-feedback cycle in bowerbirds differs from that of conventional birds in that it turns on the increased reproductive success conferred upon males able to create more elaborate and attractive bowers, rather than increased niche width. Evolution selects for ever-smarter birds.

and attempt to use this insight to alter conspecific behavior—or, as skeptics refer to it, mind reading and mind control. Whether or not the special role of subjective appreciation in bowerbird mate choice has resulted in a higher degree of social intelligence, there seems no doubt that the general intelligence of the group is as high or higher than anything we have encountered so far. The recursive cycles of selection for a set of cognitive building abilities and aesthetic refinement are part of the same sort of positive-feedback loop that may have led to the evolution of the human mind.

CHAPTER 9

Civil Engineering

LIKE REPTILES AND AMPHIBIANS, mammals rarely build structures. Except for the monotremes—the duck-billed platypus and five species of spiny anteaters—female mammals automatically provide a high degree of prenatal protection. The womb (or for marsupials such as kangaroos, koalas, and opossums, a pouch) keeps the developing young warm and relatively safe, and the female remains mobile. When mammals do build, their main impetus is protection from the elements or predators, especially for the young.

Once the young are born, a nest or den might be an advantage, but mammals generally lack either the ecological need or the morphological tools for much beyond digging, though when they do need to excavate or tunnel, mammals can move astonishing quantities of earth. Consider nine of the eleven major orders: armadillos (including sloths and anteaters) have huge blunt claws for hanging or digging; building a nest with such equipment would be like trying to write with a hammer. Or take the rabbits, whose paws are designed for rapid redeployment rather than the manipulation of objects and material. Moles and shrews are specialized diggers; their spade-like paws also preclude fine work.

Bats, though they fly and thus have a birdlike set of niches, retire to caves and similar refuges for safety. Carnivores frequently take advantage of dens, but their paws and mouths are designed for running, ripping, and killing instead of building. Whales would seem to have no use for nests, which is just as well given their short, stiff flippers. Antelope are equipped with hooves for running, and the other order with hooves, horses and the rest of the odd-toed ungulates, is equally inept at grasping and placing construction elements. The elephants could use their trunks for relatively fine construction, but their lifestyle requires no sort of shelter.

The other two orders have more scope. Among the primates, chimpanzees and humans have hands equipped with opposable thumbs that allow careful work with potential building materials. True, the chimpanzee thumb lacks muscles for rotation, making it a blunt instrument compared to our own. But then the only thing chimps seem to build are crude sleeping platforms in trees, not unlike the platforms built by magpie geese; they bend stems back over each other until they create a base.

The other mammalian "hands" are to be found in rodents. The rapid, delicate, and precise handling of seeds by squirrels is an obvious example of what this order is capable of. The real question is which species have lifestyles for which building something other than a tunnel would be useful. Squirrels, of course, build nests high in trees and use them for sleeping, rearing the young, and overwintering. Harvest mice shred grass and weave spherical nests that, except for the torn and chewed lining, could be the work of birds. But the ability to use those hands for building has been fully exploited in only two species of rodent: the American and European beavers.

Lodges

At sixty pounds or more, beavers are one of the largest of all rodents. They make their living eating the cambium and adjacent layers of tree trunks and branches. This is the ring of live tissue just under the bark, and it contains three very desirable kinds of cells. There is phloem, which consists of tubes that carry water rich in sugars and other products of photosynthesis from the leaves to sites of growth or storage. Farther into the trunk, branch, or twig is the ring of xylem, the plumbing that carries mineral-rich water up from the roots to supply the essential elements and liquid needed for growth. And between these two sets of pipes is the true cambium, which is a band of small delicate cells that reproduce to make new rings of xylem and phloem as needed. Beavers may use the leftover pile of indigestible bark chips as nesting material; the interior of trunks and branches is their version of lumber.

Beavers range throughout the temperate zone from Florida to northern Canada. They have been hunted almost to extinction in many regions, and so are more numerous in the northern pine and spruce forests simply as a result of inaccessibility. Their main predators now that the fad for beaver hats and coats has passed are highway road workers. Beavers and highway departments fight a constant battle over whether water should flow through culverts under roads or, after the beavers dam the culverts, be impounded in upstream ponds that may flood the roads after heavy rains.

Less is known about these large aquatic rodents than we might expect. Beavers are shy, mostly nocturnal, and live much of their lives hidden in lodges; even when outside, they are generally swimming, with little more than the tops of their heads or a bit of broad sleek back showing. At the same time, they ought to be one of the most interesting of all species for our purposes. They build elaborate dams and canals to control or create water flow, establish desirable water levels to ensure the safety of their communal family lodge, and generate essential transportation arteries for finding food and moving building materials. Because no two situations require the same set of designs to solve the hydraulic problems at hand, perhaps no other species outside humans has such an opportunity to display the creativity it may possess.

A mature beaver family consists of two adults and two sets of offspring, this year's kits and the previous season's yearlings. With two or three kits a year, a group may include up to eight in all. A beaver pond may be as small as a quarter of an acre or as large as a hundred acres—or indeed, there may be a vast lake that was already there, or no pond at all but instead a river the animals do not choose to alter. The pond may accommodate one family or contain up to ten active lodges. The water may be flowing or essentially still, open or covered with ice; it may be shallow or deep, and it may have woods along the edge or a quarter of a mile away. In short, there is no single typical beaver habitat; to be one of these remarkable rodents you need hardwood trees to eat, a burrow with an underwater entrance for protection from predators, and a nest above water for living. The classic beaver dam and pond are entirely optional.

Safety from predators requires a sensible use of water as a defense. A newly mated pair will search for a place to excavate a burrow. The entrance must be at least a foot under water (two feet is better), but the chamber itself needs to be above the waterline to stay dry; in fact, it has to be above the highest water level that will occur over the course of the season. Muskrats dig similar chambers, but beavers go far beyond muskrat bunkers, if circumstances permit, and erect fortress-like lodges of branches, sticks, and hardened mud.

The ideal interior room for a beaver pair seems to be six feet in diameter and perhaps eighteen inches high, with at least two entrances. But creating this kind of housing depends on the nature of the soil and topography, and it is safe to say that no two lodges are the same. The first step is almost always the burrow, which is later upgraded if possible. When the edge of the pond is steep and deep, nothing could be simpler: dig into the bank, angle up to above the waterline, and then excavate and line a chamber. As the family prospers, the chamber is enlarged upward, or a tunnel to a higher

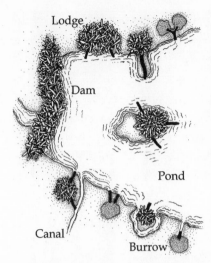

Lodge locations. A family of beavers may have multiple lodges in its home range, and these may differ greatly in location and style. In this composite sketch combining several families and ponds, there is one island lodge in the center of the pond. Moving counterclockwise from the top, there is a pair of burrows sharing a single tunnel, a lodge at the end of an artificial peninsula (created to make space for an entrance tunnel), a conventional bank lodge, a lodge built along a canal, a burrow with two entrance tunnels, a bank lodge surrounded by a moat, and a burrow at the end of a canal.

one begun. At this point, some beavers begin piling sticks up on the shore in anticipation of where their expanding chamber will break through to the surface. These are branches that have been stripped of their food layer and are intended only for construction work. Once the excavation reaches the air, they pile on more sticks to create an interlocking roof, exactly along the lines of the stick platforms of birds; the beavers use mud to seal and strengthen the structure, leaving just enough open for ventilation.

But what if the bank is not deep and steep anywhere on the pond, or at least anywhere far enough away from another lodge? One solution is to dig a deep canal into the sloping bank to create a steep slope, and then start work from under the water there. Another is to construct a mounded peninsula out into the pond, and then dig an access tunnel up into it. Or the family can build the lodge on a natural or artificial island. For an island of their own, they must bring in large foundation stones until they have a base just below the waterline. Now come sticks and mud for the floor, branches and mud for the roof, and then sticks, shavings, and grasses for the lining.

Island lodge. Island lodges may be built on a natural island, a shallow spot in the pond, or on an artificial island created by the beavers, who can bring stones from considerable distances to build a platform.

Some beavers even convert their middle-class shore lodges into island retreats by excavating a moat five feet wide around them. Despite all this investment, however, beavers take an extremely laid-back view of things. Infrared cameras inserted into lodges reveal that the beaver family coexists peaceably with a myriad of uninvited guests, including muskrats, mice, water voles, and large flying insects.

The most parsimonious interpretation of this amazing variability in construction is that beavers understand what is needed—their goal—and then come up with a strategy. From an array of innate digging and woodworking techniques they orchestrate a combination that permits them to attain their end. This Tier–6 picture is reinforced and extended by several other observations. The first is that beavers don't need to dig a burrow; families have been found in isolated caves as far south as Florida, where the local topography renders the home safe from predators and the climate makes a fully enclosed lodge unnecessary. The second is that even if they dig they don't have to *build* anything. Beavers often tunnel into human habitations adjacent to streams or ponds and set up house there. (They do, however, like to bring in sticks to create a series of runways to make moving around easier for a creature with short legs.) Finally, beavers have been observed upgrading natural brush bowers at the edge of the water into lodges, and adding a tunnel at the end, thus reversing the normal sequence of building.

This extreme flexibility in housing behavior is impressive, and seems to provide good evidence for a degree of understanding and planning that goes beyond anything we have yet seen, even in bowerbirds. There is opportunity for learning, to be sure, but it is restricted: Kits in general do not watch their own homes being erected, nor those of other families, though they do see the structure from the inside and outside, and they observe its frequent renovations.

If there is only a limited chance to learn, a skeptic seeking to deny the role of innovation and insight is in trouble. The alternative interpretation is that the behavior is largely instinctive, the result of innate programming. But that explanation requires so many conditionally independent behavioral pathways, each triggered by the appropriate set of recognition circuits (most never finding expression in the current or perhaps several recent generations), that the resulting circuit diagram seems absurd. Imagination, an ability to plan, and a ready willingness to learn from experience seem the most realistic combination of cognitive faculties to generate this aspect of the beaver's life. And this is just the burrow.

DAMS

Early naturalists thought two species of beaver lived in North America: dam beavers and bank beavers. The bank species was thought to resemble a large muskrat in behavior, content to live in a burrow or a lodge and unable to build dams. In fact, dams are primarily a strategy for dealing with the variations in water level that occur each year. Specifically, if the water level falls in the summer and autumn, as is the pattern in most of the country, then the lodge entrance may be exposed. With a dam to stabilize water level, homes will be much safer. Along deep rivers, where the "bank beavers" are to be found, the problem almost never arises. But these beavers know perfectly well how to build dams, and do so if the need arises, as may occur if they have to relocate after felling and consuming all the trees in their neighborhood.

The first serious study of American beavers was undertaken in the 1860s by L. H. Morgan, the amateur naturalist and anthropologist whose main claim to fame is his discovery that some American Indian societies are matrilineal. Morgan was a lawyer for a railroad company that was building tracks through the wilderness on the southwest shore of Lake Superior to service developing mines. The lines crossed two streams in the six-by-eight-mile area he studied, one feeding into Lake Superior and the other into Lake Michigan. On these two streams and their tributary creeks he found hundreds of dams and mapped them meticulously, including sixty-three that were more than fifty feet long. These various dams flooded between a quarter of an acre and sixty acres, the largest pond being the result of a 500-foot dam (the record is held by a 2,200-foot dam on the Jefferson River in Montana).

With the help of building crews eager to destroy inconvenient impoundments, Morgan was able to gather some fairly precise information on dam structure. One unfortunate dam, for instance, was 261 feet long, 6.5 feet high, and 18 feet thick at the base. He was able to calculate that the structure contained about seven thousand cubic feet of material, weighing perhaps a hundred thousand pounds. It incorporated trunks, limbs, stones weighing up to twenty pounds, branches, and enormous amounts of mud. Though three feet was the most common height, other dams were as much as

Beaver dam. The dam curves into the upstream direction, which allows some of the force of the water to be transferred to the bank. There is a basic (but highly variable) foundation not visible from the surface; this foundation structure is covered with smaller-gauge branches, sticks, and mud to stop the flow of water.

twelve feet tall. Those built across wide shallow streams allowed a metered amount of water to percolate through; others that closed off narrow streams with high banks had carefully crafted spillways.

Morgan thought at first that some sort of dam-building mania led beavers to create structures that were longer than they needed to be, and caused them to add intermediate dams along a stream where no pond could be created, and thus no lodge built. But as the years went on, he realized that the extra length served to reinforce the dam at its thinnest spots, and that the supernumerary dams had one of two functions. Those with spillways were built to about two feet above the level of the water they held back, with the spillway cut down to what naturally became the normal water level. That the overflow channel was a full two feet lower than the top of the dam is very unusual in beaver dams; six inches is typical for a conventional structure. These deep-spillway dams served to moderate water flow during the spring rains, keeping the extra water from draining through the shallow spillway of the main dam all at once; this allowed the impounded water in the upstream dam to rise for a time, and then drain gradually off back to the normal level. A well-designed dam balances the depth and width of the spillway to achieve a specific set of flow-limiting and pressure-mitigation goals. The other low dams, which tend to be far more numerous, simply provide a set of stairstep pools for beavers to use while carrying material up and down the stream. These dams have clear wear marks where beaver after beaver has slid down or climbed up over, probably carrying a limb in its mouth.

Morgan observed one other interesting trend: The beavers generally built the dams in his study area by cutting a log and laying it on the bottom of the stream across the direction of flow. Then they piled on more logs, using stones to help hold them in place, and sealed this porous foundation with added limbs and packed mud—an aquatic adumbration of the robin's strategy of sealing the inner layer of twigs in the cup-shaped platform of the nest. But one particularly annoying dam, which his workmen cut through at least ten times—and which the local beavers persistently repaired until they finally gave up and moved away—turned out to have been built by aligning the initial large logs *with* the stream rather than across it. Although rare in Michigan, this go-with-the-flow technique is common among beavers in France. Morgan concluded that there must be at least two framework designs available—across the current and with it.

Hardy researchers studying beavers in other places have uncovered several common but quite different techniques of dam construction. One approach is to drive sharpened stakes vertically into the stream bed and then begin adding thick straight branches horizontally between upstream and downstream stakes. As the growing structure begins to take the force of the water, the beavers reinforce the verticals with Y-shaped limbs, fitting the vertical into the crotch and wedging the other end into the downstream bottom. If the

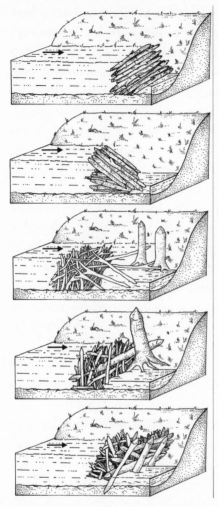

Beaver dam designs. Several different dam framework strategies are known. One option is to lay the trunks across the stream; this probably works best when the flow rate is low, and thus the limbs will not be carried away. Another possibility is to align the trunks with the flow, anchoring some into the stream. A third is to begin with a loose set of trunks placed vertically into the bed and anchored in place with branches from downstream. These braces must themselves be held in place, perhaps with stones brought to the work site. Trunks and branches are then laid across the stream with the verticals to keep them from washing away. A fourth option is to begin with one or more heavy horizontal elements felled in place, anchor them with vertical trunks, and then to fill in with more horizontal limbs. A fifth is to begin with heavy-duty verticals, followed by horizontal elements upstream, then more verticals, then another set of horizontals along with branches worked in as opportunity allows.

beaver finds it difficult to get the brace into the floor of the stream, it will bring one or more heavy stones to help. They also sometimes brace verticals against objects on the shore.

Another *modus operandi* is to fell a tree across the stream and count on its mass to hold it in place for the moment. Then sharpened stakes are driven into the floor of the creek or river downstream to hold the crosspiece against the current. The beavers add more vertical stakes upstream, and weave horizontal branches between the two. Then they stick in branches and mud to stop the flow of water. Given that few instances of dam building have been studied, odds are that beavers use more than just these few general designs for frameworks.

One thing Morgan could not have known is that when human dams became common (dams for farm ponds, mill ponds, the control of water drainage, and so on), beavers encountering these structures, built in a variety of unbeaverlike shapes and from materials unknown to them, recognized them as dams and simply modified them as necessary, raising or lowering the level, adjusting the size of the spillway, and so on. As with lodges, beavers are not driven to build dams if humans have provided a reasonable substitute. Goals rather than means seem to be their focus. Beyond the obvious mental tools that beavers seem to flaunt, such as cognitive maps and concepts, they display a remarkable degree of Tier–6 goal directedness and an ability to innovate.

Although it is a truism to say that no two dams are the same, and that each appears to reflect local needs and contingencies, the huge regional variation in styles may also reflect cultural traditions. Since the young have two years to learn something about their art from the adults, beavers might have an opportunity to pick up local customs. What proportion of youngsters are able to observe the construction of the framework of a dam rather just than its outer plastering, however, is not clear. Most youngsters grow up in territories where the major dams are already in place, and some in local habitats that do not require dams—places unlike those in which the youngsters may later set up housekeeping. Some researchers argue that the varying approaches to dam frameworks are evident in the lodges themselves, as though one basic idea were being applied to both problems. If this is true, the unending renovation of the lodges might be a source of cultural inspiration.

No matter how the dams are built, they play a crucial role when cold weather comes. Burrow and lodge entrances are built deep so that ice will never trap the family. A large pond that freezes over its entire surface can make things difficult for air-breathing animals that must then return to the burrow or lodge every few minutes for oxygen. If the ice is thin they can break open breathing holes, but for thick ice and cold weather, some families adopt a different strategy: They open a channel in their dam just below the ice, allowing a few inches of their precious pond water to drain out. The resulting gap between water and ice provides a layer of air across the entire pond.

This willingness to damage their dam contrasts greatly with the usual pattern of beavers, which is one of repair, repair, repair. By the very nature of the organic materials used—wood and mud, for the most part—the structure must be constantly strengthened and mended. For years, researchers assumed that the sight and sound of running water must trigger a mindless application of twigs and mud, but this cannot be so. For one thing, spillways are there on purpose, and are never filled in despite the almost constant sound of leaking water they can present. For another, attempts to play back the sounds of water trickling through the dam attract only mild interest (and, eventually, an attempt to encase the annoying speaker in mud); a real leak is tended to instantly. Percolation leaks through the dam produce noise and flow on the downstream side, but, except for catastrophic damage, they are always dealt with from under water on the upstream face. Trying to fit beavers into any kind of conventional stimulus-response scheme, whether it involves lodges, dams, or any of the other things they do, generally fails.

CANALS

Beavers require hardwood trees; they eat the thin nutritious layers and use the rest for building. Awkward on land, beavers prefer to

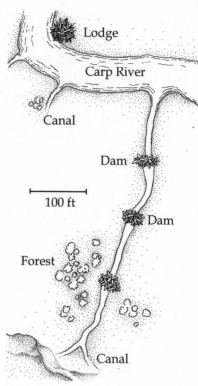

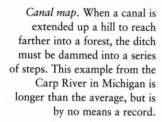

Canal map. When a canal is extended up a hill to reach farther into a forest, the ditch must be dammed into a series of steps. This example from the Carp River in Michigan is longer than the average, but is by no means a record.

commute by water, and to float their wood from where it has been cut to where it is needed. Since even a hefty branch on dry land many feet from the stream presents a serious transport problem, a large family may run through the supply of suitable trees along the edge pretty quickly. Building a dam floods more area and puts many trees within easy range.

In time, though, even the pond may not feed some growing families, and often the group must move elsewhere. But where the ground is flat enough, they adopt another stratagem, first discovered by Morgan. The beavers excavate canals into the forest and fell trees along their edges, eat the cambium, and float the branches back to the pond. From three to five feet wide and three feet deep, the canals are striking affairs, running as straight and neatly as the contour lines of the region allow. The water depth in one of their ditches begins at about thirty inches where it exits the pond and gradually declines to about eighteen inches as the ground rises. If this section of channel doesn't take the canal far enough, the beavers construct a dam and continue with a ditch on the other side that starts at the full thirty-inch depth. Morgan saw several of these channels equipped with three dams, and there seems no reason to think this is the limit.

The canals are as long as they need to be, up to six hundred feet in Morgan's area; other workers have found waterways three times this length. Sometimes these artificial creeks branch into two ditches after a couple hundred feet, each arm reaching another one or two hundred feet into widely separated patches of forest. Many heavily used channels even have small burrows built into them as refuges during the harvesting. But some canals are used for ordinary navigation; Morgan found instances in which the meandering S-curves of a river had been connected to permit cutoffs and so allow the builders more efficient movement up and downstream along a far shorter route than the river itself took. One of these connections even had a lodge built on it, an addition that supplemented the already staggering variability and flexibility of this ploy; as usual, no two were alike.

CHALLENGES

A number of experiments, planned and unintentional, tell us something more about the mind of the beaver. In one, two individuals were confined for study in an enclosure equipped with a swimming pool. This artificial pond measured roughly twelve by eight feet, its depth being about four and a half feet. The water level was not up to the rim of the pool, though, so the underemployed beavers set about correcting the shortfall. They located the outlet pipe, discovered that it had several three-quarter-inch drain holes, and gnawed twigs into sharp points to plug them. After the humans in charge kept removing the plugs so that the pool's circulation system could function, the beavers began adding increasing quantities of mud and leaves to the twigs. Beavers intensely dislike not being able to control leaks.

Another researcher attempted to keep beavers in a small park, complete with stream and pond. He provided adequate food, and tried to keep them away from the decorative shade trees by encasing the trunks in protective wire fencing. He buried the wire securely, and also tied it to the limbs overhead. One night, the beavers assembled a ramp of sticks and mud, climbed up, and cut a tree down. Others followed on almost every night. The investigator tried putting a piece of bread on a three-foot pole; since the animals were able to generalize to this quite different situation, they constructed foraging ramps. Poles without bread were never touched.

This same researcher attempted what any number of highway engineers have tried, namely, to run a mesh-covered pipe from a culvert far enough upstream or down to frustrate the dam builders. The local beavers almost always find upstream pipes and plug them, then build a dam that takes advantage of the culvert's support. We say "almost" because it seems possible to make the pipe so long and large, and cover it with three successive layers of metal mesh of widely different sizes, with suitable gaps between them, that most (but not all) beavers would give up and go elsewhere. As for downstream, the beavers simply set up their dams at the next likely site beyond the spot where the pipe opens out. They quickly figure it out when dummy pipes with artificial flows have been placed nearer the culvert as decoys; if their dam doesn't raise the water level, they start another where it will, and even use the materials from the first dam to build it.

A telling unplanned experiment occurred at a dam where observers had been watching a family of beavers for several years. The individual animals were all well known, as was their experience in building. One day the dam was vandalized by local youths, and a deep channel levered out. Water poured through, and the pond began to drain. The beavers, being late risers, did not emerge from their lodge until hours afterward, but immediately flew into action. However, the unprecedented nature of the calamity rendered the usual repair strategies useless because even large branches were swept away by the rushing water.

Before the beavers' appearance on the scene, the researchers had already been experimenting with their own repairs, but to no avail. But one of the human well-wishers had lugged up and tossed in several large stones. Like the Pullman cars dumped into the inadvertent breakout of the Colorado River through irrigation gates into the Salton Basin of California in 1905, the problem dwarfed the desperate attempts at a fill-it-in solution. But when the eldest beaver discovered these unexpected rocks, he began gathering fresh vegetation to cram between them. This tactic not only slowly sealed the stones but limited the fall in water level to a sufficient degree that the dam could then be rebuilt. But beavers have never before been known to use fresh vegetation in building. Nor do they ever repair dams from the downstream side, where in this emergency some members of this family sought a solution to the unprecedented problem. What here was luck, and what was insight? Was the successful repair an example of desperate innovation, or was it an accident?

The animals' behavior the next day was in some ways even more interesting. The eldest beaver emerged from the lodge and immediately, without a glance at the dam across the pond, removed one of the largest logs from atop the lodge. He swam with it directly to the damaged dam and wedged it into the remaining gap. One by one the rest of the family exited the lodge, looked about, and then brought a lodge limb across the pond. The first builder must have decided to bring and use the huge lodge branch before he emerged;

the others presumably saw what he was doing and followed suit. Beavers never remove material from lodges. At first sight, this appears to be premeditated planning and innovation.

Although beavers show an architectural flexibility and cleverness unprecedented outside our species, we must not forget that they share with us some deeply foolish or thoughtless moments. For instance, two beavers may work at the same time to fell a tree, but they may well gnaw at completely different heights from the ground. It's more like parallel play. Or when a tree is cut through and falls, but is held up by a neighboring tree, few beavers ever think of cutting down the adjacent tree as well, and so have two meals rather than zero. Their cognitive powers do have limitations. And then there is our own ignorance. Not only is our knowledge of their natural history woefully inadequate for such a surprising species, but no one has tried them in more familiar tests for concept formation and problem solving. (They have been tested with "puzzle boxes," the psychologist's equivalent of the interlocking twisted-wire posers found in bars. Unlike most animals, they seem to have that optimal combination of brains, patience, manual dexterity, and motivation necessary to work through the series of locks and fasteners to gain access to the interior of the box.)

Tier–5 network mapping allows an animal to orchestrate innate and learned behavioral elements to achieve innate goals; it makes possible innovation in the context of solving particular problems, and variation is almost inevitable when there are alternative routes to the same end. There is no doubt that beavers are capable of this; indeed, this view of their behavior seems limiting and inadequate. Tier–6 concept mapping opens the way for an individual to use concepts and abstract reasoning to solve problems, perhaps through insight. Almost everything about the actions of these rodents suggests that they employ concepts and reasoning to power their behavior, with insight emerging when they encounter especially difficult challenges. Their social interactions indicate a Social–2 level of network mapping to allow the coördination of individual actions to serve the family goal. As with bowerbirds, predation and food are not normally major problems; they have the time to obsess on their architectural creations, maintaining, renovating, and elaborating their dams, lodges, and canals. Their cognitive abilities have broadened their range of habitats, and that increased range has selected for yet more flexibility and creativity in dealing with the challenges that face them.

What can the architectural behavior of beavers and other animals, and any of their correlated problem-solving abilities, tell us about how our minds evolved and work? What is the source of what we call thinking, planning, imagination, and innovation? To what extent did the evolution of cognition depend especially upon the building of physical structures? Do the architectural achievements of other species support Darwin's nearly axiomatic assumption that cognition is a continuously evolved trait, subject to natural selection, and that it differs only in degree rather than in kind between species? And where building requires mental processes beyond instinct and programmed learning, are those abilities specialized and compartmentalized, as so many human mental skills are said to be? These are the questions we will turn to in the next chapter.

CHAPTER 10

Building and
the Human Mind

MOST ANIMALS FACE a common set of challenges: they must eat, survive, and reproduce. These simple words belie the complexity that distinguishes one species from another. Eating, depending on the species' niche, can involve finding food or just being born on it; it may require outwitting wary prey or merely consuming immobile leaves or detritus. Survival itself is a contest with predators, parasites, the physical elements, and chance, each depending on the habitat and lifestyle to which the species evolved. Reproduction usually includes finding or perhaps actively attracting a mate, and, for many, feeding or guarding the young.

Mental activity is, by its nature, private; what goes on in the brain has to be inferred. In tracing the evolution of cognitive strategies, the most tangible evidence is found among animals that build—in what they build and how they build it. Their structures are used for all the essential roles upon which natural selection works: prey capture, defense from predators, environmental control, mate attraction, and rearing and protecting offspring. The mental tools brought to bear include local and home-range maps, perspective shifts, network planning, and concept manipulation.

These abilities seem to have evolved independently in several different groups, but always apparently in about the same order, and to serve analogous ends.

No matter how flexible and innovative an animal seems to be, however, we have seen that that the general strategy of natural selection is to rely on and elaborate upon instinctive behavioral roots. Thus even in the most advanced species, individuals orchestrate innate elements (sign stimuli, motor programs, and drive), learned recognition, and self-conditioned behavior to produce goal-directed actions. This seems a reasonable approach, one almost inevitable given the way evolution works. Nevertheless, many researchers take the attitude that instinct and higher cognitive abilities are incompatible; they believe that a species must depend on one or the other of these mutually exclusive behavioral approaches. If this is true, and they cannot coexist, then the evolution of mental complexity we have been tracing in animal architecture is irrelevant to understanding the human mind. Is there any substance to this picture of our species as a special creation, unique among animals?

Instinct and Humans

The anti-Darwinian view that humans are different in kind rather than degree from other animals is a powerful conceit, but it does not stand up to scrutiny. It's true that our species relies heavily on culture; our attempts to teach and train infants begin at birth. This drive to intervene in a child's development, combined with the widespread belief among parents of young children that instruction and enrichment are powerful and effective tools, means that looking for convincing evidence of innate mechanisms unaffected by enculturation is tricky. But the list of possibilities is long; every early behavior that is similar across cultures is an obvious candidate, as well as many that appear later. Smiling, for instance, and the ability to recognize smiles, are good examples. Numerous experimental opportunities and empirical demonstrations prove that these abilities are innate.

One intriguing example of the continuing influence of sign stimuli in humans comes from our ultimate cultural feat, language learning. As a species-specific trick it is very impressive, and yet all normal children, bright and dull, acquire language, just as all little brown bats develop echolocation skills. However, as with bats and sonar, language learning does not require reinforcement and error correction. Children are driven to learn how to communicate, to build vocabularies, and to figure out the grammar of the tongue being spoken around them. Compare this to mastering the techniques of addition and subtraction, far easier tasks conceptually. No child learns simple math spontaneously, and some never acquire the ability at all. Reinforcement and error correction in arithmetic are essential. There is simply no innate drive to learn how to manipulate numbers beyond basic counting.

In addition to a powerful motivating drive, language learning benefits from built-in sign stimuli recognition systems that help parse the infant's acoustic world into useful categories. Human speech has fewer than three dozen consonants; this is probably a result of the constraints of our mouth and throat design, and thus we have a limited repertoire of potential sounds. Many consonant sounds are produced by similar but distinct vocal gestures. Try saying "ba," "pa," and "ma." All three depend on so-called plosive releases of air by the lips. Now try "da," "ta," and "na." Here the release of air is by the tongue placed just behind the front teeth on the alveolar (AV) ridge. Now try "ga" and "ka," which are generated farther back with the back of the tongue against the roof of the mouth.

These eight consonants sound very different to our ears. Now try saying "da" with the tongue on the teeth or behind the AV ridge; you will hear the same sound until your misplaced tongue reaches a

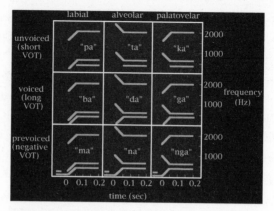

Nine stop consonants. The nine consonants made with a complete stoppage of airflow differ in where the blockage is created (lips, alveolar ridge, or the soft palate) and whether the vowel sound follows, precedes, or accompanies the stoppage (unvoiced, prevoiced, or voiced). This diagram shows the tones produced over the brief time required. The onset time of the lowest frequency is the key to voicing.

certain point, and then you will distinguish a different consonant. Or here is another example: *da*, *ta*, and *na* are made at the AV ridge, but differ when the vowel tone in the sound is generated—the part of the sound that actually engages the vocal cords. This critical moment in sound production is called the voice-onset time (VOT). For *na*, for instance, sound from the vocal cords *precedes* the release of air (a negative VOT); with *da*, they occur together; and in *ta*, the vowel follows the plosive gesture.

These are not culturally acquired distinctions. Babies just old enough to control their heads recognize and categorize the consonants in the same way. The tests depend on the short attention spans of infants and the astonishing ease with which they become bored. When a voice synthesizer alternates between consonant sounds with different VOTs that fall within the same category— negative VOT, positive VOT, or no VOT—the child's attention wanders. But if the two sounds straddle a boundary, the youngster becomes dramatically more alert. What is more, some consonants are missing from any given language; in standard English, for instance, the negative-VOT made at the back of the mouth—*nga*—is absent, though it is present in many other tongues. Children of English-speaking parents can distinguish this sound readily for several years, though adults generally lose the ability and categorize it as a *ga*. This use-it-or-lose-it phenomenon is the reason many nonnative speakers of English are unable to hear or reproduce our English *th* sound. They were not exposed to and engaged in using this consonant combination early enough.

Consonant recognition is not the only innate help we have with language. Vowel sounds, too, are processed automatically, though the boundaries are culturally defined. There are dedicated areas in the brain, generally on the left side, specific for encoding and decoding language. Higher frequencies in speech are sent to the other hemisphere to extract any emotional affect. A powerful set of circuits is specialized for deducing the rules of the local grammar; this same neural machine is able to create a grammar—a new language *de novo*—when a suitable model does not exist. Communities of deaf children will devise a language of their own that adheres to a basic set of familiar grammatical rules and contains the usual parts of speech; children reared in a group where the adults speak a variety of tongues and communicate with one another in a crude pidgin, will also create a new, mixed "creole" language. Concept formation, too, occurs automatically, allowing words to be categorized and stored in sets. In short, our greatest cognitive achievement would be impossible without the hardwired preparation of sign stimuli and dedicated processing circuits.

Another problem in appreciating the role of instinct in a fully conscious species is that the drives, sign stimuli, and innate motor programs that make life possible occur unconsciously, and thus go unnoticed. These behaviors just seem too automatic and natural to attract attention, and yet our comfort and even survival depend on

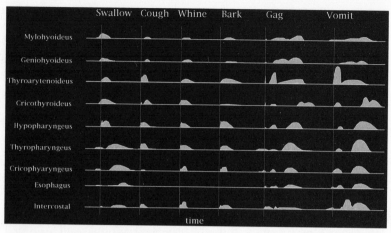

Motor programs. The nine muscles involved in six behaviors measured in dogs are shown, beginning with a muscle in the back of the mouth (mylohyoideus) and running down to the diaphragm (intercostal). For each behavior, the strength and timing of muscle contraction is shown. For swallowing, the muscles tighten in sequence from the mouth to the stomach, forcing the food down. For vomiting, the behavior starts with a gag, and then the lower six muscles tighten in a sequence that forces material up the esophagus. Each of the six behaviors is a motor program; the neural circuit for orchestrating these muscles is present from birth.

them. Orchestrated in different ways the muscles of the mouth, throat, esophagus, and diaphragm can produce a cough, generate swallowing, or produce gagging, an emergency reaction that prevents choking. The same group of muscles can work together to expel dangerous substances through vomiting. A newborn's cry is also produced by innate motor programs, as are whining, sneezing, and yawning. Walking and a variety of other more challenging behaviors are built from innate sensory and motor components.

Even learning can depend on innate drives, sign stimuli, and response elements, just as it does in other animals. The biases underlying our learning are so natural and unconscious that they rarely catch our attention. Phobias provide a case in point. Many people think of these special fears as examples of irrational learning, but they probably evolved as sensible responses to what once were real-world dangers. For instance, many animals have a special program for rapid food-avoidance learning: when a novel food makes them ill, they avoid it in the future. Thus, if rats are fed a new flavor of chow and then are made sick by being irradiated a short while later, they will avoid that flavor or odor in the future. Unlike most conditioning, only one trial is required; in the real world, one may be one too many. This same survival module is present in humans.

Certain more specific phobias, such as a fear of snakes and spiders, are very common. What key experiences (if any) are needed to trigger these phobias are not obvious, but it's easy to make the argument that they were useful when our ancestors lived in tropical jungles. The idea that they represent some rational response to real dangers is ludicrous: almost no one in the Western world dies of snake or spider bites. Tellingly, phobias do not exist for serious modern threats such as cigarettes, fatty foods, electrical outlets, or cars. (Fear of flying is probably a combination of height and confinement phobias.) These responses are not learned from a timid parent; they are bred in the bone. Every time we find something surprisingly easy or difficult to learn, we should wonder whether innate biases lie behind that cognitive interaction with the world.

We are a species equipped with its own evolutionary legacy of specialized recognition, motivation, response, and learning circuits. At least some of our behavior is robotically automatic. And yet we can still think during a yawn or a cough or while swallowing, so the presence of innate behaviors does not preclude the existence of higher cognitive processes in an animal. Their coexistence allows an animal to put much of its behavior on autopilot so that it can focus on any cognitively important tasks that arise. In theory, at least, we cannot rule out the possibility that a goose reaching for a grapefruit to roll is marveling at its crazy compulsion, much as we sometimes do at the inappropriate or inconvenient timing of a yawn, or an inability to turn down that doughnut we know rationally that we shouldn't eat. Except that we have a language that allows us to talk about it, there is no reason to think that some of our higher cognitive processing is not analogous to that of other creatures, and every reason to assume that it may have evolved in the same general way.

UNDERSTANDING

In essence, complex nervous systems exist to make sense of the world. The simplest behaviors link a stimulus to a response, S→R, as when a burned finger signals the arm to retract. Interpolation makes possible responses at intermediate sites for intermediate stimulus locations, as when an itch on an arm causes the other arm to bring fingers to the exact place to scratch. This is the first of many steps in which mapping increases the efficiency of behavior while simplifying the underlying wiring.

Conditioning, whether classical or operant, depends on circuits that extract probabilistic cause-and-effect correlations, substituting new stimuli to use in triggering old responses, or new responses to deal with familiar stimuli. In this way, a chickadee can learn to recognize sunflower seeds and develop the behavior for opening them efficiently. Local area maps allow a wide variety of animals to keep track of the relative location of nearby objects; thus they can reference their building or navigation behavior to unpredictable contingencies. Concept formation and manipulation confer new mental leverage for dealing with the world and its challenges. Each cognitive advance increases flexibility and behavioral potential, but to what extent does the animal "participate" in the experience?

The nervous system operates as though it understands cause and effect, but the creature in which these circuits reside need not comprehend anything. As we have seen, the evidence for some degree of genuine understanding appears with flexibility and innovation. For instance, when an animal can repair unlikely damage to something it has built, the simplest interpretation is that it has some kind of picture of the goal or the structure of the finished product. It sets about restoring the edifice to its former state. We see this at a simple level in caddisflies; social insects are even more flexible. Many birds can also repair damage, most impressively the weaverbirds and bowerbirds.

An ability to skip unnecessary steps, to take advantage of or compensate for unusual contingencies, to find alternative solutions to a problem, and to use novel materials may suggest more than a picture; in these situations, animals may have some understanding of the goal, the needs to be met. This is nowhere better illustrated than in beavers, though some insects and birds seem to be doing the same thing. In some sense, the ability to skip steps in a process, or to repair damage in a flexible way, are acts of extrapolation, of seeing the consequences of actions before performing them. This was the original meaning of cognitive mapping; it is the ability to choose and assemble behavioral elements in a new order, to formulate a plan.

Planning—network mapping of behavioral choices—is a big step beyond what is now commonly called a cognitive map. Home-range mapping, as we know it, is the more limited capacity to formulate new routes based on remembered landmarks or other cues. As we have seen, route planning is evident in at least some spiders and insects, as well as in birds and mammals. This particular kind of planning could be computational; there is no absolute need for understanding. But for network mapping, with its element of flexible mental extrapolation, an understanding of the goal seems essential.

If understanding requires a minimum of extrapolation and orchestration of behavioral units, then the animal must know something about the cause-and-effect relationships between materials, physical forces, and its own actions. For instance, to build with unsupported mud overhead is very different from building with the same material on a platform of sticks underneath. Wet mud has properties different from those of drier mud, and sandy mud is different from clay-rich mud, and so on. Sticks won't stay in place in some locations or orientations without being wedged or supported, and, unless they are flat, stones tend to roll down. Where it has been studied, the pattern of improvement with practice seems clear in at least birds, but is it the animal's understanding that is improving? Where does this essential knowledge of materials and physics come from?

Our immediate guess might be that experience teaches the birds all they need to know. And yet consider bowerbirds, which make a platform before trying to insert the vertical sticks that will form the walls of their alleys. Are they born knowing about gravity (in the sense that they know that vertical sticks have to be supported), or do they learn about it, discovering that unsupported sticks fall? Or do they not need to know anything about gravity because the vertical-insertion behavior is a matter of rote programming? Since most building behavior (out of the water, anyway) involves dealing with gravity, how do animals cope with this most basic cause-and-effect element in the world around them, and what sort of understanding does it involve?

The most easily interpreted work on this question comes from a set of ad hoc experiments on our closest relatives almost a century ago. A German zoologist, Wolfgang Köhler, was returning in 1917 from an extended collecting trip in Africa for the Berlin zoo. World War I erupted, and he had to land his animals on the Canary Islands until the end of hostilities. There he set up a large enclosure for his chimpanzees, and after a time began giving them various problems to solve—generally challenging them to retrieve bananas placed out of reach. He observed what he took to be planning, a controversial idea at the time. Paul Schiller's experiments on chimps some two decades later help fill out our picture of what lies behind the behavior of the animals.

The most famous problem involved simply hanging the bananas high overhead. Various chimps discovered that they could reach the food by stacking crates, or climbing poles, or standing on a crate and

Köhler's chimps. One of the chimps is knocking down the fruit with a stick, which it used after stacking two boxes underneath the food.

knocking the fruit down with a stick. From the films Köhler made, it's hard not to be struck by how incompetent they are at stacking boxes or balancing a pole before climbing. Their actions show a childish impulsiveness, an absence of a more detached perspective on the challenge. But the impressive thing to keep in mind is that the chimps seem to understand they need to begin by bringing a tool—a box or a pole or a stick—to the food; surely this is not a part of their natural innate or learned repertoire. Or is it? Köhler reported that his animals "studied" the problems, and then fetched a box or other tool with no trial and error. Schiller worried about the role of experience: Köhler had only begun testing his animals after they had spent weeks in the enclosure with the crates and other artifacts. Perhaps they had learned to solve analogous challenges in the meantime.

Schiller worked at an American primate facility on chimpanzees that had been reared in captivity under impoverished social and physical conditions. He was able to show that although training the chimps was ineffective (and perhaps even counterproductive, as Köhler, too, had reported), letting them have time to play with the crates, sticks, and poles was essential for later success. To create a plan involving the orchestration of these tools (formerly their toys), each animal needed first to understand the possibilities of the various artifacts. Since chimps seem to have a humanlike drive to experiment with novel objects, no reward was necessary. The chimps had been stacking boxes, and then climbing up and striking at the air overhead before Schiller suspended any food.

The importance of a drive to play cannot be overstated. Without undirected experimentation with objects and with the animal's own body, no very complex plan is possible. Only species with extended adolescence or an undemanding niche have a good chance of learning this way. An active childhood usually requires a social structure in which the young are taken care of. At least among higher animals, the species that show the clearest signs of planning and insight are those that indulge most obviously in play. Young ravens, for instance, engage in seemingly pointless aerial acrobatics, flying upside down, tossing rocks for others to catch, experimenting with daredevil dives, and harassing eagles.

INNATE PHYSICS?

Another set of less well-known observations by Köhler and Schiller is important to the question of an animal's understanding of simple physical reality. Stacking, for example, was inefficient because the chimps did not seem to grasp, even with experience, three basic facts. First, the bottom of the box needs to be on a flat surface; they never removed even readily transportable rocks from underneath the objective, and so the crate rocked crazily whenever the animal attempted to climb onto it. Second, a box at a higher level needs to be more or less centered above the lower one. A chimp standing on top of a disorderly stack of crates will cause everything to tumble down. Finally, the boxes need to be stacked side upon side; an upper box on one of its diagonal edges is obviously not stable. We might assume that, gymnastically talented as they are, the chimps simply enjoyed the chaos and challenge of tippy boxes. Perhaps they were not hungry, and this was just play. But the evidence is clear: The animals were strongly motivated to reach the bananas, and they became angry and frustrated with their failures. They wanted to achieve a goal, had formulated a plan for doing so, but most were unable to carry out their strategy efficiently even with practice, though some were much better than others in this regard.

The chimps' understanding of poles was also sketchy. Unlike a bowerbird that inserts a stick vertically into a platform, Köhler's primates seemed to expect that the pole would remain in place without support. Given a ladder, they refused to support it against a wall. Both Köhler and Schiller guessed that the problem was not that the animals failed to understand physics but that their innate expectations of how things work—chimpanzee physics—is quite different from ours. The chimps, for instance, would place crates against walls at shoulder height and show evident surprise that the boxes then fell to the ground. This was true whether the animals were new to the toys or had been allowed weeks of experience with them. Similarly, crates on their edges were expected to stay that way, not roll over onto a side.

They wondered whether chimps were born with an innate physics in which the ground could be assumed to be soft or covered with vegetation, thus providing an inherent element of support. Perhaps in this forest-primate world, things an animal could move and stand on (such as rocks) were usually rounded, and low-density objects generally adhered to or could be impaled on trees. Vines and other objects to be climbed might have been mostly supported from over-head or firmly rooted in the ground. Could our superior manage-ment of the same problems be the result of a different innate physics, one adapted to the open plains of our immediate ancestors, or do we come by a better genuine understanding of reality based on the strength of raw brain power? Or perhaps it's not a matter of chimps having a different innate picture of cause and effect, but in-stead an inability to deduce rules about the way things work and the forces involved from observing the consequences of their experi-mentation. What makes humans different?

Studying how human physical intuition develops is not easy. Some of the best work has been done on babies as they track the movement of objects. The eyes follow visual stimuli, and the various reactions of a baby such as surprise, continued motion of the eyes, and so on, tell us something about what the infant expects the object it is watching to do. The results suggest innate or very early predictions, with sub-sequent changes in these intuitions that probably depend on learning (though maturation cannot be entirely ruled out). For example, a toy is moved across the infant's field of view, and is tracked by the baby's eyes. In a second test, the toy is moved but now disappears behind a small curtain directly ahead. The eyes continue following the appar-ent straight-line trajectory of the object, even when nothing emerges from behind the curtain. With age, the baby shows surprise—widened eyes—when the toy disappears and fails to emerge, and will look back at the curtain. Initially, a target moving in a circle is also a source of confusion, but as an infant grows older, it is able to extrap-olate movement along this kind of trajectory as well.

The critical point here is that human infants seem to be born either assuming that objects move in straight lines, or are very much predis-posed to jump to this conclusion; apparently they must learn about other paths later. This linear-trajectory prejudice was evident in Aris-totelian physics, the "common-sense" picture of the world that dom-inated thinking for thousands of years. We are born ignorant of parabolas and gravitational constants. In fact, children and untutored adults, like countless scholars before Galileo, believe that objects move in straight lines. The actual curved track of a thrown ball (left) was assumed to be compounded of two straight lines (right):

As early as they can be tested (about three months of age), infants show an understanding of a major consequence of gravity: unsupported objects will fall. Perhaps they have learned this, but the stimuli in the tests greatly stylize and abstract the phenomenon and also provide many superficially similar control conditions. If learning is involved, as is entirely possible, it is remarkably quick and conceptual. Experts in child development are divided over whether this and numerous other instances of early knowledge represent innate expectations or innate guidance to learn certain physical concepts. In either event, all agree that the human mind is prepared for knowledge in a way that the brain of a chimpanzee is not.

Until we learn to expect something else, we interpret reality according to our innate physical rules, or rules we rapidly and selectively infer. We should not be surprised if the default expectations of arboreal, aerial, and aquatic creatures differ substantially from our own. Their perspective and planning will be greatly impacted; extrapolations may be very different; the relationship between a builder and its materials will depend enormously on this basic difference in physical context.

At the same time, an active drive to understand the forces that underlie reality is essential. Simply compiling a list of S→R observations will not in itself give the mind the ability to extrapolate in a predictive way. A striking feature of many animals, chimps included, is their apparent apathy on this score. A mania for computing arcs of movement is what allows a border collie to become an expert Frisbee catcher, while most other breeds lose track of it. Collies and certain other breeds also come to understand what pointing and gaze-direction mean. (This is a cognitive capacity, also evident in tame subordinate wolves, that has been bred for in collies; most other kinds of dogs don't get past staring at the fingertips and eyes.) Humans seem to be born believing that certain forces underlie reality, forces that must be discovered. When the phenomenon in question is largely a result of chance, our species seems driven to invent a supernatural force to account for it.

Logic

For an animal to use planning while building, renovating, or repairing—for it to take separate behavioral elements and experiences and recombine them to accomplish a goal—requires an ability to recognize logical connections and imagine the consequences of specific actions implemented in particular orders. This feat of imagination and the expectations it generates can also involve insight. We've seen apparent examples of this in nest repair and material use, as well as dam, canal, and lodge design in beavers. But can we do more than infer the existence of animal logic? The clearest experimentally controlled instance of the phenomenon comes from problem solving in ravens.

This avian version of Köhler's banana tests was devised by Bernd Heinrich, who used hand-reared birds confined to a large outdoor aviary. The birds had ample opportunity to play, at least to the extent possible in the spacious flying cage. Heinrich suspended pieces of meat on strings from horizontal branches in the aviary. Initially, the ravens flew up from the ground, grabbed the tethered meat in their beaks, and attempted to fly off. They learned almost at once that this was a useless and risky approach, and they sat instead looking longingly down at the food. Then, without preamble, one of the individuals flew to the branch, reached down with its beak,

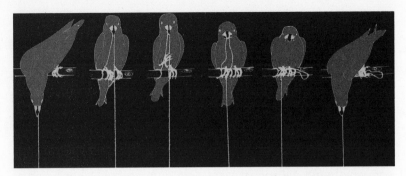

Heinrich's ravens. The first raven solved the meat-on-a-string problem by pulling up the string, stepping on it to keep it from dropping back, and then pulling up the next bit, and so on.

pulled up some string, stood on the length it had lifted, pulled up more string, stepped on it, and so on until it had the meat.

Just as with the initial solutions of the banana problem by Köhler's chimps, there was no real trial and error; the animal went to the right place and implemented a plausible solution. The problem had already been solved in the bird's mind. Other ravens devised similar ways of getting at the meat; one, for instance, grasped a length of the string and walked sideways on the perch, stepped on it and grabbed another length, walked farther to the side, and so on. One bird never cracked the puzzle; interestingly enough, this was the one raven that also failed to comprehend that it should not fly away with meat tied by a string to the branch.

To counter the criticism of skeptics, Heinrich also hung rocks from the same branch; his ravens pulled up only the strings with meat tied on. He then tied fine transparent fishing line to the suspended rocks and meat, pulling the objects to the side so that each hung directly below the spot where the other item's string was fastened to the branch. The birds showed that they understood the cause-and-effect relationship between the string and the things it connected: they pulled only on those attached to the meat. This is the kind of logic and understanding a builder needs to use network planning or concepts to create, alter, or repair a structure with the kind of flexibility and context-sensitive variability we have observed.

The cognitive interaction between the builder, its material, and the physical realities of its world raises an intriguing question: do nest builders, bowerbirds, and beavers treat individual elements—twigs, for instance—as extensions of themselves once they have grasped a building component? By an extension, we mean a physical element that reacts to the forces of the world (such as gravity and wind) until the animal takes hold of it, after which it becomes for conceptual purposes a temporary part of the animal itself, and a tool for its own insertion or placement, only to become a separate entity again once released? Does it become a temporary part of the body map in the way tools and building components sometimes do for us? Could this ability to shift perspective underlie the ability to predict the behavior and utility of objects and so contribute to the flexibility of building and repair that characterizes the most impressive animal architects?

Tools

Let's return to building and the inside-the-animal view, and try to work out how the innovative planning so evident in the architectural undertakings of many species might depend on treating building materials as tools. As we have seen, the steps of cognitive development seem to involve mapping with increasing degrees of externalization and abstractness. These mental resources are also available for other kinds of behavioral planning, of course. In our own species we depend on cognitive and network mapping, as well as the manipulation of concepts, for just about everything from fixing breakfast in the morning to getting changed for bed at night.

At first glance, building behavior seems irrelevant to most humans. In fact, our equivalent of building is primarily the creation of artifacts. Knitting and loading a dishwasher have more in common with the nest and dam construction of birds and beavers than cutting wood and nailing joists. Building is, in essence, what happens at the interface between our fingers and other moving parts and the inanimate things in our world that we choose to redesign, redeploy, or renovate. As birds use their claws and beaks, and beavers employ strong teeth and hand-like paws, we manipulate the objects around us, using them as either components or tools (or more often both). Once a familiar component is in our hand, it can become part of us conceptually, an object for which the behavioral possibilities can be estimated on the basis of past experience and then extrapolated.

The argument has often been made that only humans use tools, or rather, only humans are capable of *real* tool use. But flexible and innovative use of building components is simply tool use in one particularly common and obvious context—construction. When materials are conceptual tools for a species, the use of one for multiple purposes, such as nest building and foraging, could reveal the nature and degree of flexibility in the animal's mind. One striking example of what a cognitively advanced animal can do by redirecting elements of its building behavior is the New Caledonian crow.

These highly social birds are endemic to—that is, native to and found nowhere else than—tropical New Caledonia and an adjacent island, at the southeast edge of the Coral Sea. The nest-building behavior is typically crow-like: the birds break sticks from trees to construct a sturdy platform, then create a cup lined with strips of vegetation. Since New Caledonian crows are omnivorous, eating fruits and nuts as well as insects, the need for behavioral flexibility is built into their diet. Lacking the long and pointed bills of woodpeckers, they have nevertheless evolved into the woodpecker niche by creating two kinds of tool to get at hidden food. The simplest, a long shaft with a hook at the end, looks like a miniature harpoon that the crows fashion from secondary twigs wrenched from a larger primary branch. The bit of wood that comes off the tree with the

Crow tools. New Caledonian crows commonly make two kinds of tools.
(A) Hook tools are made by removing a twig from a branch, and then stripping
it of leaves and side twigs. The scar where the twig pulled away from the branch
provides the hook for capturing insect larvae in holes and under bark. Shown
here are a twig torn from the branch, but yet to be stripped, and a stripped and
trimmed tool. (B) Step tools are made by cutting out a long edge of screw pine
(pandanus) leaf. Screw pines are succulents with an upper growth form similar
to yucca and aloe, but a base reminiscent of a palm tree: thick, fleshy, sword-like
leaves grow up and out from a short, rough trunk. The leaf edge has a series of
small barbs. The stepping of the cut leaves is such that the barbs are generally
curved back toward the wider end of the tool (which the bird holds in its beak).
The narrow end is used to probe in holes and under bark. Here we see a leaf
with an almost complete tool cut into it, and at the bottom are four finished
step-cut tools that indicate the range of individual variation.

twig, appropriately trimmed, becomes the barb that hooks the prey.
The birds cut the shaft itself to about eight inches long, then strip it
of leaves and bark.

The other class of tool is even more impressive, and is built in
quite a different way. The crows fabricate these so-called step-cut
implements from the stiff leaves of screw pines—a leaf that has
curved barbs along its outer edge. The crow cuts the tool to three
widths: a terminal section about an eighth of an inch wide and two
inches long, an intermediate section perhaps a quarter of an inch
wide and again two inches long, and a base roughly half an inch
wide and four inches long. During construction, the crow positions
itself so that the barbs on the finished product are oriented back toward the wide base. Then, using the base as a handle, the bird em-

ploys this formidable weapon to extricate insects and their larvae from trees.

So what is going on in the brain of the crow as it fashions these tools? A few obvious scenarios come to mind. Construction and use might be wholly or partially innate. Another possibility is that the behavior could be entirely cultural: the basic utility of a twig might have been discovered by some especially canny or lucky crow, passed on by observation to its colleagues, and in a like manner the various steps of manufacture would have been the product of trial and error. But for a nonhuman animal to observe another animal performing a behavior and then copy it is almost unknown in nature, and has never been documented in birds.

A third possible explanation is that the behavior might depend entirely on insight-based innovation in the face of a novel problem. This seems at first glance unlikely: even though the crow is legendarily among the cleverest of birds, no other kind of crow is known to do anything like this. Manufacturing and using woodpecker tools is apparently unique to a single species of corvids. And if New Caledonian crows are specially gifted with this sort of creativity, why are there only two tool types? Surely there must be dozens of other ways of making such implements. Moreover, why should nearly all the step tools have three segments? Why not two, or four, or none at all?

As usual, we are presented with necessarily incomplete observations, only a partial grasp of the ontogeny, and a set of equally likely or unlikely hypotheses. Our understanding of the way New Caledonian crows use tools is an intriguing mixture of knowledge and ignorance, and we know just enough from field observations to reveal a unique behavior. But luckily in this instance there is a rare opportunity for controlled experimentation because researchers are beginning to rear these crows in the lab. In one test, lab-reared juveniles were offered screw-pine leaves for the first time. Each made a plausible, though somewhat incompetent, imitation of the tools built in the wild. It is clear that parts of the material choice and fabrication behavior do not need to be learned from scratch. So in using the tools they are innately inclined to make, do the birds understand the cause and effect relationships involved?

Two New Caledonian crows, a wild-caught juvenile female and a much older male from a zoo, were brought into the lab for a tool-choice test. Both had had uncontrolled experience before arriving, though it's unlikely that they had picked up anything directly relevant to the problem they were offered. A year before the experiments, however, the female had encountered a pipe cleaner and bent it. In the test, meat was placed in a metal bucket at the bottom of a transparent tube. Two alternative tools were available—a straight piece of wire, and one with a hook. To solve the problem, a bird needed to select the wire with the hook, hold the tool at its straight end, use the hook to catch the handle of the bucket, and lift the food container out. This totally unnatural behavior should occur by chance only at a very low rate.

As so often happens, animals do things spontaneously; behaviors appear that researchers have not allowed for in their experimental designs. In this instance, the male seized the hooked wire and went off with it, away from the tube. The female then took the remaining (straight) wire and, after discovering that it was useless as offered, proceeded to bend it to create a hook; she then used the hook to extricate the metal pail. The test was immediately reorganized to use just one bird at a time, with only the single piece of straight wire available. In nine of ten trials, the female bent the wire and recovered the bucket; the male, by contrast, never bent the wire, though he did somehow succeed in getting the food once with a straight wire. The male never understood the six-second trick of bending the wire, though he had ample opportunity to observe the female's successes.

This test tells us several revealing things. Most important, the crow solved a highly artificial lab-based problem using a material not available in the wild, which it manipulated in a unique way. We don't know what the female had already seen and done on New

Crow problem solving. This female New Caledonian crow
spontaneously solved the problem of how to extract the food bucket
from the plastic cylinder by bending a piece of wire to create a hook;
the wire was then used as a tool to lift the bucket out by its handle.

Caledonia before she was captured, but it seems obvious that in the
lab she solved a novel problem through insight—by understanding
the goal and imagining the potential of the wire. But she always
began by trying to use the unbent wire, spending a fairly consistent
fifteen seconds on this pointless enterprise; she would then take the
wire elsewhere and, after twenty seconds or so, create the bent-wire
tool. Bending was quite variable, both in technique and eventual
shape; there was no operant conditioning evident.

This leaves us unsure about how much innate help the female
crow had available. Perhaps she knew innately that proper food is
invariably at the bottom of holes, and that she is supposed to get at
it by holding something long and thin in her beak. Although that
much seems plausible, and wholly in accord with the innate help
seen in learning-dependent building behavior, bending the wire is
new; no amount of innate programming, observational learning, or
conditioning before capture could have contributed much of anything to her performance. It involves insight and planning, and yet
is curiously stereotyped: from eleven to nineteen seconds trying with

the straight wire, twenty seconds or so to find a place to bend it to an angle of (very) roughly 74 degrees, six seconds (plus or minus two seconds) to create the tool one way or another, and then back to the food tube. Is the bird reinventing the wheel each time, or is she simply compelled to try with the straight wire first before she "renovates" it? Perhaps the bending makes use of behavioral components associated with breaking twigs for making nest elements and harpoon tools—subroutines and drives that come into play once the animal discovers that the long thin object she has discovered does not work straight out of the box.

Cognition and the Human Mind

The evolutionary scenario we have traced for the development of cognitive abilities in animals has focused on building behavior. Positive feedback drives selection for greater mental ability; new abilities lead to more opportunities, and thus to an advantage for more cognitive equipment and intellectual flexibility. New mental abilities have emerged from previous ones in a logical order, beginning with simple body mapping and ending with the mental manipulation of tools in the broadest sense. An increased facility in using any given degree of mapping has often been favored by selection, but it is the steps from one kind of mapping to the next more externalized and abstract kind that constitute the most obvious directional changes. At each step, the morphological equipment of the species is critical to what is possible—the structure of onboard tools such as beak and claws, or teeth and paws, or hands and a vocal apparatus capable of generating multiple consonants and vowels.

This is not to say that selection must always be at work, or even operate in this one "upward" mental direction. The experimental psychologist Marian Breland pointed out that every animal is about as smart as it needs to be. In her view, there is rarely selection for unnecessary intelligence, and evolution can root out useless cognitive skills when a change in niche makes them obsolete. If humans are smarter in many ways, it is because our ancestors needed to be, and our morphology allowed it. Our mental development may be greater and spread out over more years of adolescence, but it is still curiously specialized and stereotyped, probably as a consequence of our inheritance.

What might have led to the kinds of minds we now possess? Our basic niche opened up perhaps 10 to 15 million years ago. Continental spreading created the Rift Valley in Africa, slowly splitting open a tropical forest inhabited by our arboreal and essentially vegetarian ancestors as well as other primates. The widening valley created an ever-drier habitat, first of deciduous forest and then of open savanna. Ours was the first species of primate to move into this new pair of habitats, with their many vacant niches. A world dominated by climbing in the third dimension became one of moving about on a plane. Food was far more dispersed. Selection favored walking, freeing arms and hands for more tool use. Tool use, in turn, would have selected for greater creativity and flexibility in manipulating things, as well as for a gradual repositioning of the muscles that make our thumbs, no longer needed for locomotion in the trees, so good at handling objects.

Selection must have also have operated on our mapping skills once our ancestors had to range so widely. As animal prey entered the diet, the cognitive skills for using mental maps to plot strategy and guess what a target organism will do must have been strongly favored. Based on comparisons between chimpanzees and the few hunter/gatherer cultures that survive, the social structure must have also changed, and with it the nature of our social intelligence. Chimps forage mainly for fruit, an easy target requiring no great degree of coöperation or strategizing. The social system is based on (mostly physical) male dominance. Even so, some hints of what would later evolve are present in our closest relative. Chimpanzees engage in very limited group hunting of prey (generally other primates), territorial defense, the strategic elimination of competing groups, social networking, and an ability to guess at the intentions of other members of the group. Other traits less acceptable in human societies are also evident, including cannibalism and mental illness.

The enormous increase in relative brain volume in the line leading to humans tells us that the positive-feedback loop for cognitive potential has reached extraordinary strength. Relative brain volume is most often measured as the ratio of brain mass to body weight. The assumption is that as adult body size gets larger, the number of sensory neurons and the circuits needed to analyze them, as well as the number of muscles and motor neurons required to control them, rises. There is no a priori way to know at what ratio these two numbers should change, given the probability of efficiencies of scale. But by plotting the two values against each other for a wide range of warm-blooded animals, the rate of typical increase can be inferred: brain mass goes up only about 40 percent as quickly as body weight. This means we can predict the brain size of a species from its adult weight, and note which ratios seem unusually high or low.

Primate brain-volume ratio is roughly double that of other groups of mammals. The most likely reason is the huge range of motion of the limbs combined with relatively fine control of the digits. To take full advantage of the possibilities generated by the morphology for grasping and swinging by the arms, primate brains must accommodate the neural machinery to monitor and control the sensors and muscles; with this wiring comes an unprecedented ability to orchestrate the manipulation of objects. Four million years ago, with our ancestors (*Australopithecus*) beginning to live full time in the savanna, the brain/body ratio was above the warm-blooded mean, but not too much more so than for typical primates, including chimpanzees.

From *Australopithecus* to the present the brain/body ratio has increased dramatically. Human brains are now two and a half times larger than would be expected even for a primate; we are the far-

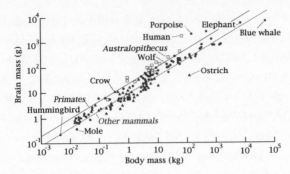

Brain–body regression. Plotting the ratio of brain weight to body mass for warm-blooded animals yields a fairly consistent regression line. Primates (open squares) tend to fall above this line, which suggests they have more cortical material than is strictly necessary. Humans are farthest above the line, ostriches farthest below. Values for aquatic mammals are hard to interpret since their body mass is supported by water rather than muscles that require neural control, and much of their weight is devoted to heavy insulation (blubber).

thest off the warm-blooded average of all species. This huge increase preceded, and possibly helped make possible, the migration of some of our ancestors out of the savanna and into other parts of Africa, as well as Asia and Europe.

We have good reason to think that the true ratio of human processing power to the primate norm is still higher; many of our cognitive tasks, unlike those of other species, have been lateralized into one hemisphere or the other in an apparent effort to save processing space. Moreover, our cortex, compared to that of other species, is highly folded, cramming still more processing area into a limited space. Much of the increase, whatever the exact figure is, must be devoted specifically to higher cognitive tasks, which multiplies this differential still further. Some of this capacity is devoted to social intelligence—quite a lot of it if we include language in this category, as seems reasonable. Most of the rest must center on the manipulation of tools, including building elements and concepts.

For our species, the two kinds of intelligence, generating an upward spiral of cognitive potential both separately and together, created our world-dominating technology. No one knows what factors selected for this runaway explosion in mental ability. It seems to have predated language, but it appeared after our transition into the savanna. Having the hands free to build and manipulate, to make real the things we might imagine, looks like the most important change. The mastery of fire, a gradual process beginning with opportunistic use 1.8 million years ago and culminating in universal cooking 200 thousand years ago, reduced our need for food by about half. This freed up enormous amounts of spare time for creative use, and allowed our species to spread first out of the tropics, and then out of Africa.

Competition, however, was necessary to drive the evolution from this point; this means competition for resources within the social group (the reward being in offspring), or competition between groups perhaps including physical confrontation; also competition between species for food—sometimes with serious carnivores, such as lions and hyenas, who have invested more in physical rather than cognitive specializations for hunting. The ultimate competition would have been the race against chance and the elements by a relatively defenseless two-legged primate. Since clubs and spears are tools that can evolve culturally far faster than claws and muscles—equipment that depends on a few minute's network mapping rather than generations of selection—our ancestors may not have been unprotected for long.

But for the selection to have gone on for more than 3 million years there must have been multiple causes, or one cause that can plausibly have stayed with us the entire time. In the latter, the pressure behind the repeated cycles of the social and conventional intelligence loops has to have been other humans, with whom we share the same niches. Across a species increasingly dependent on its mind and the technology that its nervous system generates, evo-

Notable dates in human evolution (MY=millions of years before present)

2.6 MY	*Australopithecus africanus;* first stone tools
2.3 MY	
2.0 MY	*Homo habilis*
1.7 MY	Opportunistic use of fire; Ice Age begins
1.4 MY	Ice Age ends
1.2 MY	Gradual beginning of most recent Ice Age
1.0 MY	*Homo erectus*
0.8 MY	
0.6 MY	
0.5 MY	Universal control of fire; humans begin moving out of tropics
0.45 MY	Relative brain size begins a rapid increase
0.4 MY	
0.35 MY	
0.3 MY	*Homo neanderthalensis*
0.25 MY	
0.2 MY	*Homo sapiens;* cooking universal; end of most recent Ice Age
0.15 MY	Humans begin to spread out of Africa
0.1 MY	Neanderthals appear; relative brain size stabilizes; language universal
0.05 MY	Humans reach North America and Australia
0.01 MY	Domestication of plants and animals

lution will reward the wit to innovate, the ability to manipulate perspective rapidly, a capacity to understand forces and extrapolate consequences, brains that expect a world of cause and effect, and mental machinery that maps materials and tools into extensions of the hands. And human hands are the most flexible manipulators on the planet.

Once the competition began to turn as much on social and conventional intellect as strength, the cognitive escalations that led to wider niches and new species in nest-building species became, for ours, an engine that created new incidental possibilities as fast as it exploited the old ones. Lewis Wolpert defines consciousness as an internal model of what we are doing and an ability to decide how to behave. By this measure, many species must be at least dimly sentient. Through selection for more and more cognitive flexibility in handling materials and treating them as tools, consciousness at this level must have developed most often in the widespread context of building. The key factor that seems to be most exaggerated in our species is not simply consciousness in Wolpert's sense but instead intelligence and imagination—network mapping and concept use. The limitations in the last many centuries have been more often ones of materials than imagination, of figuring out how to create new tools instead of how to use them.

Our present world, alas, is one in which most of us might perish with the loss of some key element of technology—the internal combustion engine, for instance, or electricity. The deep impulse to manipulate the world around us has wedded us to those artifacts as surely as the robins' survival is tied to the species' nest. We are the ultimate inheritors of a drive hundreds of millions of years old to build, and thus take charge of our immediate surroundings. For better or for worse, this architectural drive eventually created the kind of mind we now possess.

READINGS

CHAPTER 1 READINGS:

J. H. Fabre (1915). *Hunting Wasps* (New York: Dodd Mead).

J. L. Gould (1982). *Ethology: The Mechanisms and Evolution of Behavior* (New York: W. W. Norton).

J. L. Gould (2002). Instincts to learn, in R. Gallistel (ed.), *Stevens' Handbook of Experimental Psychology*, 3rd ed., 239–257 (New York: Wiley).

J. L. Gould and C. G. Gould (1999). *The Animal Mind* (New York: W. H. Freeman).

K. Z. Lorenz and N. Tinbergen (1938). Taxis and instinctive action in the egg-retrieving behavior of the greylag goose (English translations in C. H. Schiller (ed.) [1957], *Instinctive Behavior*, 176–208 [New York: International Universities Press]; and K. Z. Lorenz [1970], *Studies in Animal and Human Behavior*, vol. I, 328–342 [Cambridge, MA: Harvard University Press]).

D. A. Spalding (1973). Instinct: with original observations on young animals. *MacMillans Magazine 27*, 282–293 (reprinted in J. L. Gould [1982], *Ethology: The Mechanisms and Evolution of Behavior*, A1–A10 [New York: W. W. Norton]).

N. Tinbergen (1938). On the orientation of digger wasps (English translation in N. Tinbergen [1972], *The Animal in Its World*, vol. 1 [Cambridge, MA: Harvard University Press]).

CHAPTER 2 READINGS:

W. H. Bristowe (1958). *The World of Spiders* (London: Collins).

J. Dembowski (1933). Über die Plastizität der tierischen Handlungen, Beobachtungen und Versuch an *Mollana* Larven. *Zoologische Jahrbücher 53*, 261–312.

J. H. Fabre (1916). *The Life of the Caterpillar* (New York: Dodd Mead).

K. von Frisch (1974). *Animal Architecture* (New York: Harcourt Brace).

J. L. Gould (1986). The locale map of honey bees: do insects have cognitive maps? *Science 232*, 861–863.

D. R. Griffin (2002). *Animal Minds* (Chicago: University of Chicago Press).

M. H. Hansell (1984). *Animal Architecture and Building Behaviour* (London: Longman).

E. L. Kessel (1955). The mating activities of balloon flies. *Systematic Zoology 4*, 97–104.

W. G. van der Kloot and C. M. Williams (1953). Cocoon construction by the cecropia silkworm. *Behaviour 5*, 141–156.

W. Köhler (1922). *The Mentality of Apes* (New York: Harcourt Brace).

H. W. Levy (1978). Orb-weaving spiders and their webs. *American Scientist 66*, 734–742.

L. P. Lounibos (1975). The cocoon spinning behaviour of the Chinese oak silkworm. *Animal Behaviour 23*, 843–853.

W. Sattler (1955). Über den Netzbau der Larve von *Hydropsyche angustipennis*. *Naturwissenschaften 42*, 186–187.

W. Sattler (1965). Über Mebvorgänge bei Bautätigkeit von Trichopteren-Larven. *Zoologischer Anzeiger 175*, 378–385.

M. S. Tarsitano and R. R. Jackson (1994). Jumping spiders make predatory detours requiring movement away from prey. *Behaviour 131*, 65–73.

J. B. Wallace and D. Malas (1976). Fine structure of the capture nets of the larval Philopotamidae. *Canadian Journal of Zoology 54*, 1788–1802.

J. B. Wallace and F. F. Sherberger (1975). The larval dwelling and feeding structure of *Macronema transversum*. *Animal Behaviour 23*, 592–596.

G. B. Wiggins (1977). *Larvae of North American Caddisflies* (Toronto: University of Toronto Press).

CHAPTER 3 READINGS:

G. P. Baerends (1941). Fortpflanzungsverhalten und Orientierung der Grabwespe. *Tijdschrift voor Entomologie 84*, 68–275.

W. Beebe (1947). Scale adaptation and utilization in *Aesiocopa patulana*. *Zoologica 32*, 147–152.

W. Beebe (1953). A contribution to the life history of the euchromid moth *Aethria carnicauda*. *Zoologica 38*, 155–160.

H. J. Brockmann (1980). The control of nest depth in a digger wasp. *Animal Behaviour 28*, 426–445.

H. E. Evans and M. J. West Eberhard (1966). *The Wasps* (Ann Arbor: University of Michigan Press).

J. H. Fabre (1915). *The Hunting Wasps* (New York: Dodd Mead).

J. H. Fabre (1914). *Mason-Bees* (New York: Dodd Mead).

J. H. Fabre (1916). *More Hunting Wasps* (New York: Dodd Mead).

K. von Frisch (1974). *Animal Architecture* (New York: Harcourt Brace).

C. G. Gould (2004). *The Remarkable Life of William Beebe, Explorer and Naturalist* (Washington, DC: Island Press).

J. A. A. van Iersel and J. van dem Assem (1964). Aspects in the orientation of the digger wasp *Bembix rostrara. Animal Behaviour Supplements 1,* 145–162.

A. P. Smith (1978). An investigation of the mechanism underlying construction in the mud wasp *Paralastor* sp. *Animal Behaviour 26,* 323–240.

N. Tinbergen (1938). On the orientation of digger wasps (English translation in N. Tinbergen [1972], *The Animal in Its World,* vol 1 [Cambridge, MA: Harvard University Press]).

CHAPTER 4 READINGS:

H. E. Evans and M. J. West Eberhard (1966). *The Wasps* (Ann Arbor: University of Michigan Press).

K. von Frisch (1974). *Animal Architecture* (New York: Harcourt Brace).

J. L. Gould and C. G. Gould (1999). *The Animal Mind,* rev. ed. (New York: Freeman).

M. H. Hansell (1984). *Animal Architecture and Building Behaviour* (London: Longman).

H. Kemper and E. Döhring (1961). Zur Frage nach der Volksstärke und der Vermehungspotenz bei den sozialen Faltenwespen Deutschlands. *Zeitschrift für Angewandte Zoologie 48,* 163–197, 255–309.

W. E. Kerr, S. F. Sakagami, R. Zucchi, V. de Portugal-Araújo, and J. M. F. de Camargo (1967). Observaçóces sôbre a arquitetura dos ninhos e comportamento de algumas espécies de abelhas sem ferrão das vizinhanças de Manaus, Amazonas. *Atas do Simpósio sôbre a Biota Amazônica, Conselho Nacional de Pesquisas, Rio de Janeiro 5,* 255–309.

H. Martin and M. Lindauer (1966). Sinnesphysiologische Leistungen bein Wabendau der Honigbeine. *Zeitschrift für vergleichende Physiologie 53,* 372–404.

O. W. Richards and M. J. Richards (1951). Observations on the social wasps of South America. *Transactions of the Royal Entomological Society of London 102,* 1–170.

H. de Saussure (1853). *Monographie des guêpes sociales ou de la tribu des vespiens, ouvrage faisant suite à la monographie des guêpes solitaires* (Paris: V. Masson).

J. H. Sudd (1967). *Introduction to the Behaviour of Ants* (London: Arnold).

E. O. Wilson (1971). *The Insect Societies* (Cambridge, MA: Harvard University Press).

E. O. Wilson (1981). Communal silk spinning by larvae of *Dendromyrmex* tree ants. *Insectes Soc.* 28, 182–190.

CHAPTER 5 READINGS:

K. von Frisch (1967). *The Dance Language and Orientation of Bees* (Cambridge, MA: Harvard University Press).

K. von Frisch (1974). *Animal Architecture* (New York: Harcourt Brace).

J. L. Gould and C. G. Gould (1995). *The Honey Bee*, rev. ed. (New York: Freeman).

J. L. Gould and C. G. Gould (1999). *The Animal Mind*, rev. ed. (New York: Freeman).

M. H. Hansell (1984). *Animal Architecture and Building Behaviour* (London: Longman).

W. E. Kerr, S. F. Sakagami, R. Zucchi, V. de Portugal-Araújo, and J. M. F. de Camargo (1967). Observaçôces sôbre a arquitetura dos ninhos e comportamento de algumas espécies de abelhas sem ferrão das vizinhanças de Manaus, Amazonas. *Atas do Simpósio sôbre a Biota Amazônica, Conselho Nacional de Pesquisas, Rio de Janeiro 5*, 255–309.

M. Lindauer (1978). *Communication Among Social Bees* (Cambridge, MA: Harvard University Press).

M. Luscher (1961). Air conditioned termite nests. *Scientific American 205* (1), 138–145.

H. Martin and M. Lindauer (1966). Sinnesphysiologische Leistungen bein Wabendau der Honigbeine. *Zeitschrift für vergleichende Physiologie 53*, 372–404.

C. D. Michener (1974). *Social Behavior of Bees* (Cambridge, MA: Harvard University Press).

T. D. Seeley (1982). *Honey Bee Ecology* (Princeton: Princeton University Press).

T. D. Seeley, P. K. Visscher, and K. M. Passino (2006). Group decision making in honey bee swarms. *American Scientist 94*, 220–229.

E. O. Wilson (1971). *The Insect Societies* (Cambridge, MA: Harvard University Press).

CHAPTER 6 READINGS:

W. Dilger (1962). The behavior of lovebirds. *Scientific American 206* (1), 88–102.

H. J. Frith (1957). Experiments on the control of temperature in the mound of the mallee fowl. *CSIRO Wildlife Research 2*, 101–110.

H. J. Frith (1962). *The Mallee Fowl: The Bird That Builds an Incubator* (Sydney: Angus & Robertson).

K. von Frisch (1974). *Animal Architecture* (New York: Harcourt Brace).

P. Goodfellow (1977). *Birds as Builders* (Oxford: David & Charles).

C. G. Gould (2004). *The Remarkable Life of William Beebe* (Washington, DC: Island Press).

M. Hansell (2000). *Bird Nests and Construction Behaviour* (Cambridge: University of Cambridge Press).

H. H. Harrison (1975). *A Field Guide to Birds' Nests* (Boston: Houghton Mifflin).

R. J. Herrnstein and D. H. Loveland (1964). Complex visual concepts in the pigeon. *Science 146,* 549–551.

R. J. Herrnstein, D. H. Loveland, and C. Cable (1976). Natural concepts in pigeons. *Journal of Experimental Psychology 2,* 285–302.

C. A. Ristau (1991). Aspects of the cognitive ethology of an injury-feigning bird, the piping-plover, in C. A. Ristau (ed.), *Cognitive Ethology,* 91–126 (Hillside, NJ: Lawrence Erlbaum).

B. F. Skinner (1960). Pigeons in a Pelican. *American Psychologist 15,* 28–37.

D. Stokes (1979). *A Guide to Bird Behavior,* vol. I (Boston: Little, Brown).

D. Stokes (1983). *A Guide to Bird Behavior,* vol. II (Boston: Little, Brown).

CHAPTER 7 READINGS:

A. B. Addicott (1938). Behaviour of the bush-tit in the breeding season. *Condor 40,* 49–63.

N. E. Collias and E. C. Collias (1984). *Nest Building and Bird Behavior* (Princeton: Princeton University Press).

W. Dilger (1962). The behavior of lovebirds. *Scientific American 206* (1), 88–102.

K. von Frisch (1974). *Animal Architecture* (New York: Harcourt Brace).

P. Goodfellow (1977). *Birds as Builders* (Oxford: David & Charles).

M. Hansell (2000). *Bird Nests and Construction Behaviour* (Cambridge: University of Cambridge Press).

H. H. Harrison (1975). *A Field Guide to Birds' Nests* (Boston: Houghton Mifflin).

A. Kulczycki (1973). Nesting of members of the Corvidae in Poland. *Acta Zoology, Cracov 18,* 583–666.

P. L. Lee, D. H. Clayton, R. Griffiths, and R.D.M. Page (1996). Does behavior reflect phylogeny in the swiftlets? A test using cytochrome *b* and mitochondrial DNA sequences. *Proceedings of the National Academy of Sciences 93,* 7091–7096.

I. Rowley (1978). The use of mud in nest building—a review of the incidence and taxonomic importance. *Ostrich, Supplement 8,* 139–148.

W. Scheithauer (1967). *Hummingbirds* (London: Crowell).

B. E. Smythes (1968). *The Birds of Borneo* (Edinburgh: Oliver & Boyd).

D. Stokes (1979). *A Guide to Bird Behavior,* vol. I (Boston: Little, Brown).

D. Stokes (1983). *A Guide to Bird Behavior,* vol. II (Boston: Little, Brown).

H. O. Wagner (1955). Einfluss der poikilothermie bei kolibris auf iher Brutbiologie. *Journal für Ornithologie 96*, 361–368.

K. Zyskowski and R. O. Prum (1999). Phylogenetic analysis of the nest architecture of neotropical ovenbirds. *The Auk 116*, 891–911.

CHAPTER 8 READINGS:

G. Borgia (1985a). Bower quality, number of decorations, and mating success of male satin bowerbirds. *Animal Behaviour 33*, 266–271.

G. Borgia (1985b). Bower destruction and sexual competition in satin bowerbirds. *Behavioural Ecology and Sociobiology 18*, 91–100.

G. Borgia (1986). Sexual selection in bowerbirds. *Scientific American 254* (6), 92–100.

G. Borgia (1993). The cost of display in the nonresource-based mating system of the satin bowerbird. *American Naturalist 44*, 734–743.

G. Borgia (1995a). Why do bowerbirds build bowers? *American Scientist 83*, 542–545.

G. Borgia (1995b). Complex male display and female choice in the spotted bowerbird. *Animal Behaviour 49*, 1291–1301.

G. Borgia (1995c). Threat reduction as a cause of difference in bower architecture, bower decoration, and male display in two closely related bowerbirds. *Emu 95*, 1–12.

G. Borgia and M. A. Gore (1986). Feather stealing in the satin bowerbird. *Animal Behaviour 34*, 727–738.

G. Borgia, I. M. Kaatz, and R. Condit (1987). Flower choice and bower decoration in the satin bowerbird. *Animal Behaviour 35*, 1129–1139.

G. Borgia and U. Muller (1992). Bower destruction, decoration stealing, and female choice in the spotted bowerbird. *Emu 92*, 11–18.

G. Borgia, S. Pruett-Jones, and M. Pruett-Jones (1985). The evolution of bower-building and the assessment of male quality. *Zeitschrift für Tierpsychologie 67*, 225–236.

K. Collis and G. Borgia (1993). The costs of male display and delayed plumage maturation in the satin bowerbird. *Ethology 94*, 59–71.

N. E. Collias and E. C. Collias (1984). *Nest Building and Bird Behavior* (Princeton: Princeton University Press).

W. T. Cooper and J. M. Forshaw (1977). *The Birds of Paradise and the Bowerbirds* (Sydney: Collins).

C. Darwin (1871). *The Descent of Man and Selection in Relation to Sex* (London: Murray).

J. Diamond (1982a). Evolution of bowerbirds' bowers: animal origins of the aesthetic sense. *Nature 297*, 99–102.

J. Diamond (1982b). Rediscovery of the yellow-fronted gardener bowerbird. *Science 216*, 431–434.

J. Diamond (1986a). Biology of birds of paradise and bowerbirds. *Annual Review of Ecology and Systematics 17*, 17–37.

J. Diamond (1986b). Animal art: variation in bower decorating style among male bowerbirds. *Proceedings of the National Academy of Sciences 83*, 3042–3056.

J. Diamond (1987). Bower building and decoration by the bowerbird. *Ethology 74*, 177–204.

J. Diamond (1988). Experimental study of bower decoration using colored poker chips. *American Naturalist 131*, 631–653.

C. B. Frith, G. Borgia, and D. W. Frith (1996). Courts and courtship behavior by Archbold's bowerbird. *Ibis 138*, 204–211.

C. B. Frith and D. W. Frith (1995). Court site constancy, dispersion, male survival, and court ownership in the male tooth-billed bowerbird. *Emu 95*, 84–98.

C. B. Frith and D. W. Frith (2004). *The Bowerbirds* (Oxford: Oxford University Press).

E. T. Gilliard (1963). The evolution of bower birds. *Scientific American 209* (2), 38–46.

E. T. Gilliard (1969). *Birds of Paradise and Bower Birds* (London: Weidenfeld & Nicolson).

J. L. Gould and C. G. Gould (1996). *Sexual Selection,* rev. ed. (New York: Freeman).

D. R. Griffin (2002). *Animal Minds: Beyond Cognition to Consciousness* (Chicago: University of Chicago Press).

M. Hansell (2000). *Bird Nests and Construction Behaviour* (Cambridge: University of Cambridge Press).

S. Humphries and G. D. Ruxton (1999). Bower-building: coevolution of display traits in response to the costs of female choice? *Ecology Letters 2*, 404–413.

C. P. Hunter and P. D. Dwyer (1997). The value of objects to satin bowerbirds. *Emu 97*, 200–206.

R. Kusmierski, G. Borgia, R. H. Crozier, and B. H. Y. Chan (1993). Molecular information on bowerbird phylogeny and the evolution of exaggerated male characteristics. *Journal of Evolutionary Biology 6*, 737–752.

R. Kusmierski, G. Borgia, A. Uy, and R. H. Crozier (1997). Labile evolution of display traits in bowerbirds indicates reduced effects of phylogenetic constraint. *Proceedings of the Royal Society of London, B 264*, 307–313.

N. Lenz (1994). Mating behaviour and sexual competition in the regent bowerbird. *Emu 94*, 263–272.

A. J. Marshall (1954). *Bower Birds* (London: Oxford University Press).

S. Pruett-Jones and M. Pruett-Jones (1994). Sexual competition and courtship disruption: why do male bowerbirds destroy each other's nests? *Animal Behaviour 47*, 607–620.

R. E. Vellenga (1970). Behaviour of the male satin bower-bird at the bower. *Australian Bird Bander 8*, 3–11.

R. E. Vellenga (190). Distribution of the bowers of satin bower-birds at Leura, NSW, with notes on parental care, development, and independence of the young. *Emu 80*, 97–102.

J. Warham (1957). Notes on the display and behaviour of the great bower-bird. *Emu 57*, 73–78.

CHAPTER 9 READINGS:

K. von Frisch (1974). *Animal Architecture* (New York: Harcourt Brace Jovanovich).

D. R. Griffin (2001). *Animal Minds* (Chicago: University of Chicago Press).

L. H. Morgan (1868). *The American Beaver and His Works* (Philadelphia: Lippincott).

P. B. Richards (1983). Mechanisms and adaptation in the constructive behavior of the beaver. *Acta Zoologica Fennica 174*, 150–158.

H. Ryden (1989). *Lily Pond* (New York: William Morrow).

L. Wilsson (1971). Observations and experiments of the ethology of the European beaver. *Viltrevy: Swedish Wildlife 8*, 115–266.

CHAPTER 10 READINGS:

R. Baillargeon, L. Kotovsky, and A. Needham (1996). The acquisition of physical knowledge in infancy, in D. Sperber, D. Premack, and A. J. Premack (eds.), *Causal Cognition*, 79–116 (Oxford: Clarendon).

D. Bickerton (1983). Creole languages. *Scientific American 249* (1), 108–115.

T.G.R. Bower (1982). *Development in Infancy* (San Francisco: Freeman).

R. Corrigan and P. Denton (1996). Causal understanding as a developmental primitive. *Developmental Reviews 16*, 162–202.

P. D. Eimas et al. (1971). Speech perception in infants. *Science 171*, 303–306.

N. Geschwind (1979). Specializations of the human brain. *Scientific American 241* (3), 180–199.

J. Goodall (1986). *Chimpanzees of Gombe* (Cambridge: Harvard University Press).

J. L. Gould (1982). *Ethology: The Mechanisms and Evolution of Behavior* (New York: W. W. Norton).

J. L. Gould and C. G. Gould (1999). *The Animal Mind* (New York: Scientific American Library).

J. L. Gould and P. Marler (1987). Learning by instinct. *Scientific American 256* (1), 74–85.

D. R. Griffin (2002). *Animal Minds* (Chicago: University of Chicago Press).

B. Heinrich (1995). An experimental investigation of insight in common ravens. *The Auk 112*, 994–1003.

G. R. Hunt (1996). Manufacture and use of hook-tools by New Caledonian crows. *Nature 379*, 249–251.

H. W. Köhler (1922). *The Mentality of Apes* (New York: Harcourt Brace); translation of W. Köhler (1921) Intelligenzprüfungen an Menschenaffen (Berlin: Springer).

S. Pinker (1994). *The Language Instinct* (New York: William Morrow).

D. J. Povinelli (2000). *Folk Physics for Apes* (Oxford: Oxford University Press).

P. H. Schiller (1957). Innate motor action as a basis of learning, in C. H. Schiller (ed.), *Instinctive Behavior*, 264–287 (New York: International Universities Press).

A. S. Weir, J. Chappell, and A. Kacelnik (2002). Shaping of hooks in New Caledonian crows. *Science 297*, 981.

L. Wolpert (2003). Causal belief and the origins of technology. *Proceedings of the Royal Society of London A 361*, 1709–1719.

CREDIT LINES FOR FIGURES

Every effort has been made to find the actual rights holder for each of the photographs and drawings in this volume.

Page

x   Authors' collection.

6   From J. L. Gould and W. T. Keeton (1996). *Biological Science,* 6th ed. (New York: W. W. Norton).

8   Based on G. P. Baerends and J. P. Kruijt (1973). Stimulus selection, in R. A. Hinde and J. Stevenson-Hinde (eds.), *Constraints on Learning* (New York: Academic Press), 23–49, with permission from Elsevier.

10   Authors' collection.

14   Authors' collection.

15   Authors' collection.

23   Authors' collection.

27   Authors' collection.

29   Authors' drawing.

33   Authors' drawing.

37   From G. B. Wiggins (2004). *Caddisflies: The Underwater Architects*
(top)   (Toronto: University of Toronto Press, 2004), with the kind permission of G. B. Wiggins.

37   Drawing by Sandy Rivkin based on J. B. Wallace and F. F. Sherberger
(bottom)   (1975). The larval dwelling and feeding structure of *Macronema transversum. Animal Behaviour 23,* 592–596, with permission of Elsevier.

41   Authors' drawing.

Page

43   From M. S. Tarsitano and R. R. Jackson (1994). Jumping spiders make predatory detours requiring movement away from prey. *Behaviour 131,* 65–73, with permission of Koninklijke Brill NV.

44   Authors' collection.

47   Reprinted by permission of HarperCollins Publishers Ltd. © W. H. Bristowe (1958), from *The World of Spiders.*

49   Authors' collection.

51   From H. W. Levy (1978). Orb-weaving spiders and their webs, *American Scientist 66,* 734–742, with permission of Sigma Xi, the Scientific Research Society.

55   With permission of Peter Chew.

60   Authors' collection.

62   Authors' collection.

66   Drawing by Sandy Rivkin.

68   From A. P. Smith (1978). An investigation of the mechanism underlying construction in the mud wasp *Paralastor* sp. *Animal Behaviour 26,* 232–240, with permission of Elsevier.

70   From A. P. Smith (1978). An investigation of the mechanism underlying construction in the mud wasp *Paralastor* sp. *Animal Behaviour 26,* 232–240, with permission of Elsevier.

71   From A. P. Smith (1978). An investigation of the mechanism underlying construction in the mud wasp *Paralastor* sp. *Animal Behaviour 26,* 232–240, with permission of Elsevier. (both images)

83   Authors' collection.

88   Authors' collection.

89   Authors' collection.

90   Authors' collection.

93   From H. de Saussure (1853–1858). *Monographe des guêpes socials ou de la tribu des vespiens, ouvrage faisant suite à la monographie des guêpes solitaires* (Paris: V. Masson).

96   Courtesy of the artist, Turid Forsyth. (both images)

102   Courtesy of the artist, João M. F. de Camargo.

109   Authors' collection. (both images)

112   From J. L. Gould (1982). *Ethology* (New York: W. W. Norton). (all three images)

114   From *The Honey Bee,* © 1988 by J. L. Gould and C. G. Gould, reprinted by permission of Henry Holt and Company.

115   From *The Honey Bee,* © 1988 by J. L. Gould and C. G. Gould, reprinted by permission of Henry Holt and Company.

Page

116　Authors' collection. (A)

116　Authors' drawing. (B)

117　Authors' drawing.

119　From *The Honey Bee,* © 1988 by J. L. Gould and C. G. Gould, reprinted by permission of Henry Holt and Company.

125　From *The Honey Bee,* © 1988 by J. L. Gould and C. G. Gould, reprinted by permission of Henry Holt and Company.

131　Authors' drawing.

134　From M. Luscher (1961). Air conditioned termite nests. *Scientific American 205* (1), 138–145.

135　Authors' collection. (both images)

137　From *Animal Architecture* by Karl von Frisch and Otto von Frisch, drawings © 1974 by Turid Hölldobler, reproduced by permission of Harcourt, Inc. (A)

137　From J. Desneux (1956). Structures "atypiques" dans les nidification souterraines d'*Apicotermes lamani. Insectes Soc. 3,* 277–281, with permission of Birkhäuser Verlag. (B)

138　From *The Animal Mind,* © 1994 by J. L. Gould and C. G. Gould, reprinted by permission of Henry Holt and Company.

140　From *The Animal Mind,* © 1994 by J. L. Gould and C. G. Gould, reprinted by permission of Henry Holt and Company. Based on *Animal Architecture* by Karl von Frisch and Otto von Frisch, drawings © 1974 by Turid Hölldobler, reproduced by permission of Harcourt, Inc.

143　From E. O. Wilson (1971), *The Insect Societies* (Cambridge, MA: Harvard University Press), drawing by Turid Forsyth, with permission of the artist.

151　Authors' drawing.

155　Drawing by Sandy Rivkin based on H. J. Frith (1962). *The Mallee Fowl: The Bird that Builds an Incubator* (Sydney: Angus and Robertson).

158　Authors' collection; from W. Beebe (1918), *A Monograph of the Pheasants* (London: H. F. Whitherby).

163　Authors' collection.

166　Authors' collection.

168　Authors' collection.

175　Authors' drawing.

178　Authors' drawing.

181　Based on D. W. Winkler and F. H. Sheldon (1993). Evolution of nest construction in swallows: a molecular phylogenetic perspective. *PNAS 90,* 5705–5707, copyright 1993 National Academy of Sciences, U. S. A., with permission.

Page

238 From E. T. Gilliard (1969). *Birds of Paradise and Bower Birds* (London: Weidenfeld & Nicolson) with permission of Weidenfield & Nicolson, a division of the Orion Publishing Group.

242 From E. T. Gilliard (1969). *Birds of Paradise and Bower Birds* (London: Weidenfeld & Nicolson) with permission of Weidenfield & Nicolson, a division of the Orion Publishing Group.

244 From J. Diamond (1987). Bower building and decoration by the bowerbird. *Ethology 74*, 177–204, with permission of Blackwell Publishing.

249 Authors' drawing.

255 Drawing by Sandy Rivkin based on L. H. Morgan (1868). *The American Beaver* (Philadelphia: Lippincott).

256 From L. H. Morgan (1868). *The American Beaver* (Philadelphia: Lippincott).

259 From L. H. Morgan (1868). *The American Beaver* (Philadelphia: Lippincott).

261 Drawing by Sandy Rivkin based in part on P. B. Richard (1955). Bièvres constructeurs de barrages, *Mammalia 19*, 293–301.

264 Drawing by Sandy Rivkin based on L. H. Morgan (1868). *The American Beaver* (Philadelphia: Lippincott).

274 Based on W. Strange & J. Jenkins (1978). In R. D. Walk & H. L. Pick (eds.), *Perception and Experience* (New York: Plenum), with permission of Springer Verlag and Winifred Strange.

276 Based on R. W. Doty and J. F. Bosma (1956). An electromyographic analysis of reflex deglutination, *Journal of Neurobiology 19*, with permission of John Wiley & Sons.

281 From W. Köhler (1921). Intelligenzprüfungen an Menschenaffen (Berlin: Springer).

287 From B. Heinrich (1995). An experimental investigation of insight in common ravens. *The Auk 112*, 994–1003, with permission of the American Ornithologist's Union and *The Auk*.

290 Authors' collection. (A)

290 Drawing by Sandy Rivkin based on G. R. Hunt (2000). *Proc. Royal Society London B, 267*, 403–413, with permission of the Royal Society. (B)

293 With kind thanks to and permission of Alex Kacelnik.

297 Based on H. J. Jerison (1969). Brain evolution and dinosaur brains. *American Naturalist 103*, 575–588.

INDEX